Instruments
of
Amplification

Fun With Homemade
Tubes, Transistors, and More

Written and Illustrated by
H. P. Friedrichs

Dedicated to childhood curiosity...
May it live in all of us forever

Table Of Contents

Chapter 1
Introduction

Back in 1997, I stumbled across an old book that discussed the application of certain naturally occurring minerals to early radio receivers. Now, one might think that an electrical engineer would be weary of technical puzzles after a busy day at work, but not so. For reasons I can't explain, I have an inexhaustible interest in science, and that interest had once again been piqued. I decided that a crystal receiver project would prove an interesting and relaxing diversion.

Crystal-radio sets are standard fare as electronic projects for beginners. In recent years, they've attracted a new following among dedicated hobbyists and the nostalgic alike. As I engaged in this project, I began to collect and read books on the subject.

There, I noticed a recurring theme. Construction plans always required the purchase of certain critical components. Stated or implied, these "special" parts were thought to be too difficult to construct at home. Now here was a tacit challenge I simply couldn't pass up! I resolved to design and build a working radio receiver *entirely* from scratch, that is to say, no store-bought electrical parts allowed. I limited myself to the use of wire and common hardware store or garbage can materials.

As the various components to my radio began to develop, it occurred to me that other people might consider my work interesting and wish to replicate it. As a result, I made the crucial decision to *document* what I was doing. By the time that effort was complete, and it had been fleshed out with suitable background material, theory, and illustrations, I held in my hands a respectable body of work. It became my first book, ISBN Number 0-9671905-0-9, entitled: *The Voice of the Crystal.* (I'll call it "*VOTC.*")

If you've ever given consideration to authoring a book yourself, I would urge you, by all means, to buckle down and *do it!* Far from being a simple means to an end, the experience allowed me to meet dozens of wonderful and interesting people I'd not have met otherwise. It also taught me some useful lessons that have directly impacted the book you're about to read.

For example, I have long been troubled by the general public's apparent lack of interest in things of a scientific or technical nature. This is particularly disturbing among the young, who are the essence of our future. I mentioned this concern in *VOTC's* introduction, bemoaning a world of technical sophistication filled with people who have neither an idea nor a care as to how it all works. I was pleasantly surprised at the enthusiasm with which my first book was received, and encouraged that scientific apathy might not be as acute as I had imagined. By the time this book goes to press, there will no doubt be several *thousand* copies of *VOTC* scattered across the globe. Thus, I discovered that there were more people like me than I had first imagined, and that writing a second book would be worth my time and effort.

Another interesting lesson had to do with the presentation of the projects themselves. I had decided to fabricate *VOTC's* projects in ways more or less consistent with early radio. This meant the use of stained and sealed woods, polished brass, screw terminals, and an overall look and feel in keeping with turn-of-the-century laboratory equipment. Admittedly, these devices could have been implemented with more contemporary materials, with less care and less attention paid to their aesthetics. Yet, in choosing to build "museum pieces," I eventually recognized a benefit that was then entirely unforeseen.

I propose that one of the reasons for the public's lack of interest in science is that modern technology, for all its sophistication, tends to be sterile and uninspiring in its raw form. Take the average integrated circuit chip, for example. It may be capable of establishing communications with a satellite in Earth orbit, or executing millions of calculations per second, but its dull, gray plastic exterior says nothing about the absolute *magic* that occurs inside. In its native state, it's no more inspiring than an equally sized wad of gum.

In demonstrating the various instruments from *VOTC*, I'd anticipated interest from people with electronics backgrounds. What astonished me most, however, was the attention lavished upon them by individuals with no technical inclination whatsoever. The latter felt compelled to handle these "gizmos," turn them over in their fingers, and study their construction. The art in these machines had stoked their curiosity, and somehow inspired them to ask questions that might never have occurred to them otherwise. For this reason, I decided that any projects for a new book would be fabricated with this phenomenon in mind.

Once I had decided to produce a new book, I began to contemplate its subject matter. I had met my objectives in the *VOTC*, and the book concluded with functional radio equipment. However, that equipment was passive in nature, representative of the technology dominant during radio's early infancy.

The problem with a passive receiver is that the power of the music or speech delivered to its headphones can never be more than the amount of power collected by its antenna. Furthermore, radio signals degenerate rapidly as the distance from the transmitter to receiver is increased. Historically, there would be no "golden age" of radio until scientists and engineers had developed some way to *amplify* incoming signals. Tubes, and later transistors, led to the resulting explosion in wireless electronic communication.

The subject of amplification, then, seemed to be a natural topic for a new book. In fact, several of my readers made proposals to this effect. They suggested that I develop a series of projects featuring simple transistor or tube-based circuits. I declined because, quite frankly, there's no shortage of that type of information. If you take into account the selection of excellent modern texts, countless old books, and then add to that a plethora of Internet Web sites, there are literally *thousands* of circuits and projects from which to choose.

Still, I liked the "amplification" theme. The more I thought about it, the more I began to view it through the "build-it-from-scratch" ethic on which the *VOTC* had been based. Slowly, I began to envision a book not about amplifier circuits, but about the *amplifying devices themselves.* I would attempt to build homemade tubes, transistors, and other amplifying devices...you've got it...from scratch. This launched an amazing adventure, the conclusion of which now lies in your hands: *Instruments of Amplification.*

Thus inspired, and in many ways shaped by its predecessor, this book might be seen as a sequel to *VOTC.* A radio enthusiast can easily begin with *VOTC* projects, and then advance to the projects found here. At the same time, I prefer to think of *Instruments of Amplification* as an individual work to be judged on its own merits. The subject material here is certainly more sophisticated, and the successful completion of these projects ultimately requires more commitment in terms of time, materials, and study. The rewards, of course, are also proportionately greater.

Obviously, no book can be all things to all people. I've had to make certain assumptions about the intended audience, so that decisions could be made about the scope and content.

First, is the treatment of technical theory. Throughout this volume, you'll find accounts of basic principle and the reasoning behind the instruments I describe. Be warned, that those explanations are cursory and anything but comprehensive. My focus here is the *practical* implementation of those ideas. Therefore, it's my assumption that the reader will supplement his or her literary "diet" with as many additional books as necessary. (Warm up that library card!) There's no reason for me to rehash details that are readily found in the works of a hundred other authors.

Second, it's a bitter irony that the most powerful tool of science, namely mathematics, seems to be the greatest factor in discouraging prospective experimenters. I've purposefully avoided the excessive use of formulas and expressions, though I'll assume the reader is familiar with Ohm's law and understands the basic differences between a volt, an ohm, and an ampere. If not, any introductory book on electricity will arm you with the requisite knowledge.

You will probably notice that most of the chapters in this work conclude with a fairly extensive list of references. These references may include books, magazine articles, newsletters, and Web sites. They represent the primary source of much of the information contained here. I've taken the liberty to cite specific page numbers and to comment on the nature of the information found on those pages. Here and there, I've also included books I consider to be useful resources, even if they were not directly utilized in the preparation of my text.

Having said all this, it's probably a good idea to insert appropriate disclaimers so that we may proceed with a reasonable set of expectations. If, for example, you're a high-end audio fanatic or an electric guitarist with a taste for tube amplification, and it's your dream to manufacture plug-in replacements for your favorite "fire bottles," you'll probably be disappointed with my projects. Realistically, commercial-grade tubes require techniques and materials that are well beyond the scope of a book like this one.

On the other hand, high-quality blown-glass tubes have been produced in the workshops of a handful of dedicated enthusiasts. If this is your ultimate intent, you can do no better than to jump right into the types of projects found here. They're a great way to learn what works and what doesn't without an unreasonable investment in time and materials.

The same can be said for the transistor information found here. For all its conceptual simplicity, a useable transistor is notoriously difficult to build. Success is frightfully dependent upon the materials used in its fabrication. Modern transistor production involves techniques that simply cannot be replicated in the home workshop.

However, alternate materials and processes can result in completed devices that clearly demonstrate transistor action. Under good circumstances, they may even prove functional in practical circuits.

In the end, the value of this book lies not in the physical equipment you build. The reward is ultimately measured in the breadth and the depth of the knowledge you'll acquire. The technologies described here, like most, are based in multiple disciplines of science. If you read this book, attempt the experiments I describe, and strive to *understand* them, you'll eventually find yourself literate in the subjects of electricity, magnetism, electrostatics, vacuum practice, physics, chemistry, and much, much more. Best of all, this sort of "mad" science is pure, unadulterated fun! So, roll up your sleeves and get ready.

While I take full credit (or blame!) for the production of this book, I would be remiss in failing to mention some of the numerous people whose encouragement served to assure its ultimate completion: T. Lindsay of Lindsay Books, R. Hewes of the Xtal Set Society, K. Christiansen, J. Ward, D. Moore, M. Nelson, S. Coles, K. Sunamura, N. Steiner, J. Jenkins, and many more. A special thanks goes to J. Habermas for his effort and skill in editing and reviewing my final draft.

You have not read a book like this before.

Chapter 2
Basic Tools

Years ago, I owned an old car that had developed a problem with its turn signals. Living in an apartment at the time, I had no garage or other suitable place to work on it, and parking lot repairs, as such, were frowned upon by the apartment's management. Just the same, I could neither afford nor justify the expense of dealership service. Being handy with things of an electrical nature, I had no doubt that I could diagnose and correct the problem. I took the plunge and attacked it myself.

First, I spent a good hour playing with the light fixtures, probing with my digital multimeter for continuity and proper grounds. I traced wires through the fenders, engine compartment, and the firewall. Everything looked to be in order. In fact, I ultimately concluded that the fault lay with the turn signal switch. Admittedly, I embraced this revelation with some dread. The switch was buried deep inside the steering column, and the only way at it was to take everything apart.

First, I had to detach the steering wheel itself. Removing the plastic trim from the wheel was no problem. The nut that secured the wheel to the steering shaft yielded to a large wrench. The wheel itself required some coaxing, as the steering shaft was splined, but through patience and a bit of ingenuity, it finally popped free. To my utter surprise and absolute horror, the entire assembly exploded in a hail of mechanical parts, springs, and other widgets. For an instant, my blood ran cold at the thought that some vital component might have rocketed into low-earth orbit, vanished in the lawn, or tumbled into the storm sewers. Thankfully, at the time, I was seated in the driver seat with the doors closed. All of the escaping fragments were contained within the passenger compartment, so at least I had a fighting chance of finding them all and somehow putting that mess back together.

With the column disassembled, I quickly verified that the signal switch was indeed faulty. Repairing it was a snap. The turn signals were restored to proper operation, so I redirected my attention to the challenge of reassembling everything.

I should mention that this car was equipped with an adjustable "tilt-wheel," an attractive and convenient feature that unfortunately depends upon a complicated arrangement of bearings, springs and other parts. The intent of the extra parts is to ensure that steering is smooth, crisp, and without slop or backlash, regardless of the wheel's angle of tilt. It soon became apparent that some kind of special fixture or tool had been employed at the factory to compress the springs and hold the various pieces in alignment to permit the steering wheel to be slid onto its shaft. Needless to say, I had no such tool.

Ultimately, the answer to my problem lay in the contents of two or three old coffee cans. There, I've made a habit of collecting nuts, bolts, bits of steel, and other odds and ends. To make a long story short, I was able to *fabricate* my own tool, based on a little trial, a lot of error, and an educated guess or two. In the end, it worked well enough to allow me to reassemble the various pieces and reattach the wheel. Granted, the process took several hours, but all told, the operation was an unqualified success.

Experience has taught me that there is a preferred tool for every task. Having the proper tool for the job at hand can result in enhanced safety, significant savings in time, and a reduction in the effort required. As a bonus, the end result of that effort is generally of a much higher quality.

Yet, ironically, the lesson implicit in my story is precisely the opposite! Yes, having the "right" tools helps, but lacking those tools does not automatically doom your efforts to failure. Care, patience, and a bit of ingenuity will often go a long way toward offsetting the handicaps in your equipment and materials. You may be able to work around a problem, or arrive at your goal through alternate means. Or, as was the case in the tale above, it may be possible to *build* the very tool that you need!

I had planned to introduce, in general terms, some of the tools that are useful in conducting the experiments in this book. I wanted to make certain, however, that the technically timid and those in my audience of limited means wouldn't be discouraged or put off by discussions of numerous or expensive items. Just remember the moral of the story: The "right" tools are nice to have, but most of the time, you can get by with far less. In the chapters to follow, you'll find commentary based on the equipment I have and the techniques I've employed, though where possible, I've always tried to mention and explore alternatives.

That said, all of the work described in this book demands a minimal set of hand tools. An assortment of screwdrivers in various sizes, both flat blade and cross-point (phillips head), are a must. As well, pliers like needle-nosed, slip-joint, and side-cutters round out the basic kit. Most people own a hammer and a simple hand saw. A measuring tape and a small square helps. In my opinion, no one should be without these basic items, if for no other reason than to execute simple domestic repairs.

Once you get into project building, particularly small machines or other technical or scientific equipment, it helps to expand your inventory just a bit. The next level includes tools for cutting and shaping, like files, metal shears, hacksaw blades, and the like. Be warned that in this arena, the catchphrase: "you get what you pay for," is gospel. I tend to spend the extra dollars to purchase well-made, brand-name tools. Fit and finish are superior; these tools function more smoothly and reliably, and if properly cared for, they'll last a lifetime. Generally you won't get that degree of service from dime-store tools.

Speaking of service, you'd be amazed what you can do with an electric hand drill. Obviously, with the appropriate bits, you can put a hole into just about anything. What's less evident is that, if applied properly, a drill motor can be used to fashion small, rounded parts in much the same manner as a lathe.

For what it's worth, I recently purchased a rechargeable drill, and have become a devout fan of this technology. Aside from its obvious portability and convenience when used for outdoor projects, I've found it quite suitable for use in winding delicate coils of wire. In the past, I'd depended on mains-powered drills with some success, but I've since noticed that the trigger on my rechargeable unit offers much finer resolution and better control of the chuck speed. This is a big plus when you're winding coils in small gauges of wire, because it helps to minimize tangles and the potential for breakage.

Small drill presses, particularly those of the import variety, have become extremely inexpensive. If you have the bench space, they're a worthwhile investment. The principle advantage of the drill press is that it allows one to locate and drill holes with precision. It also helps ensure that holes are drilled perpendicular to the work surface. If you're drilling fragile materials like glass, a topic I'll touch upon in a later chapter, the drill press is almost indispensable. A hand drill will tend to wander under those circumstances, and glass articles will invariably be broken.

For my own purposes, I've found 8-inch presses just a trifle small, opting instead for a 12-inch model. I've never regretted that decision. On a side note, you can purchase mechanical adapters that are capable of converting an ordinary hand drill into a primitive drill press. I used one of these before purchasing my present drill press, and actually did a fair amount of work with it. It offers many of the advantages of a real press, while remaining affordable.

Speaking of useful tools, I can't overemphasize the convenience and utility of owning a lathe. A couple of years ago I purchased a small model-maker's lathe, having been tempted by a local deal that I couldn't pass up. The rig's a well-made Austrian import, scarcely the size of a small sewing machine. With it and a small assortment of 1/4" cutting bits, you can turn tiny parts, cut gears, bore holes, cut grooves or screw threads, and even do some light milling. It's not an overly powerful device, but it's more than capable of turning mild steels and works like a champ on brass, plastics, and hardwood. I'm not ashamed to admit that the tool's capabilities readily exceed my limited skills as a machinist. In fact, my only complaint is that it's been so much fun, I now lust after a larger, more powerful model!

In as much as this book is about electronic amplification, it stands to reason that there are certain pieces of electrical and electronic equipment necessary to perform this kind of experimentation. At a bare minimum, you'll need a decent soldering iron, and a handheld multimeter.

Your soldering iron should be a pencil-type iron, suitable for small electronic work. The gun-type iron that many people own comes in handy for certain applications, but tends to be gross overkill most of the time. Pencil irons are readily available at hardware stores, or dealers in electronics.

Basic multimeters are cheap. For less than 10 dollars, you can get an instrument that will read voltage, current, and resistance on an analog dial. More expensive models are digital, and offer convenient features like auto-ranging, capacitance, and frequency measurement ranges.

To be candid, I've been disappointed by local sources of electronics, and have turned to mail-order channels for my parts and equipment. Take a trip to your local library and browse the advertising sections of such magazines as *Nuts & Volts*, *Popular Electronics*, or one of the ham magazines like *QST* or *Practical Wireless*. If you're Web-enabled, the world is at your fingertips. There are dozens of electronic retailers and surplus houses out there.

One of the most useful pieces of electronic equipment an experimenter can have is an oscilloscope. Basically, it does no more than plot electrical waveforms on a screen, yet in terms of utility, a scope is to the engineer what a lathe is to the machinist. In the hands of a person who knows its full potential, there are endless possibilities: you can measure voltages, currents, frequency, modulation, phase angles, and characterize semiconductors. High-performance digital scopes can cost many thousands, but I've seen used analog scopes go for as little as 25 dollars. Check the online auction sites, mail-order distributors, and the Web. Flea markets are another possible source.

If you're tinkering with amplifiers, it's helpful to have a source of stable and controllable test signals. What you need is a signal generator. Like scopes, signal generators come in a variety of flavors, and in a range of prices. A basic model will generate sine waves, triangle waves, and square waves, and will allow you to adjust the frequency from DC to a couple of hundred kHz. Used, they're often inexpensive. I know of one electronics outfit that offers a stripped-down generator as a kit. It can be yours for a twenty-dollar bill. Don't be afraid of used equipment. Most of my test gear is secondhand and well-worn, but was obtained for pennies on the dollar.

Let's not overlook power supplies. There's nothing in this book that can't be driven by a suitable stack of common batteries, but I prefer to use an adjustable laboratory-style power supply. One of my favorites is a beat-up old Lamda. The model I have actually features two independent supplies in one cabinet, each adjustable from 0 to 40 volts, and each capable of supplying up to an amp. Power supplies are common projects for electronics "beginners," so schematics and instructions for building them from scratch are readily available in books and on the Web.

Vacuum tubes depend on the absence of air for their operation. If you want to play with experimental tubes, you'll need some way of drawing a vacuum. The production of a useable vacuum is a subject too complex to resolve in a single paragraph, so we'll revisit the matter in a later chapter. There, we'll discuss what's meant by the phrase "vacuum," relative degrees of vacuum, and schemes for removing air from your experimental devices. Even if you never build a single tube, there are dozens of fun and instructive experiments you can do with a glass jar devoid of air. High voltages, for example, applied to electrodes in a canning jar, can produce wonderful, blue-glowing discharges that swirl and dance under the influence of a magnet. It's like having a ghost in a bottle!

Chapter 3
Safety First

The world is an uncertain place and life itself, if not precious, is definitely precarious. Every year, thousands of ordinary people suffer injury and death simply in going about their business. In 1998, for example, more than 41,000 people died in the routine of driving from point "A" to point "B".

One wonders if they were headed home. Had they made it alive, they might have been among the 22,000 injured by washers and dryers, 42,000 injured by television sets or the 53,000 who were injured by an oven or stove. (These figures come from the National Highway Traffic Safety Administration, the U.S. Consumer Product Safety Commission, and the National Safety Council.) Life itself, it would appear, is hazardous to your health.

Interestingly, statistics suggest that the odds of drowning in your own bathtub are 100 times greater than the advertised odds of winning the super jackpot in a popular lottery game. Now, if I developed a "system" to radically improve a person's odds of winning that lottery, is there any doubt that crowds would beat on my door demanding the secret? Why is it then, that a "system" which improves the odds of surviving *life* is frequently met with indifference or a polite yawn? I'm speaking, of course, of the philosophy known as *safety*.

We live in an imperfect world. Despite well-laid plans and the best of intentions, things occasionally go wrong. Safety is about anticipating and planning for mistakes. It's about improving your odds and stacking the deck in your favor. It's not something to be overlooked or pooh-poohed. It's an important tool that smart players use to win the game of life.

This book is about hands-on science. It's about the use of tools, electricity, vacuums, chemicals, and more. While possibly unfamiliar, they shouldn't be feared, at least no more than the "hazardous" bathtub installed in your home. The key to safety is that they be treated with informed respect, and utilized with common sense. Throughout this book, I've provided ample notice where a process or technique might prove hazardous. For now, bear with me as we touch upon some general, but crucial, safety highlights.

Protect your eyes! Existing medical technology can do little to repair a tattered eye, and even less to replace it. Safety glasses provide cheap insurance, and are available in various models at hardware and home-improvement stores. Take the time to read and understand the literature that comes with protective eye wear. Make certain that you understand the capabilities of your glasses, and that they're appropriate for the work you're doing.

In most cases, you'll want protective glasses with high-impact lenses, and frames with shields to prevent particles from ricocheting into your eyes from the sides. In some cases, goggles are the better choice. They may offer a similar degree of impact protection, and because they seal against your face, they offer superior protection against splashing liquids.

Protect your ears! The mechanism that enables you to hear is exceedingly delicate. One hundred million years of evolution could not have prepared it for the cacophony that makes up the modern soundscape.

Obviously, extreme noises can result in an instant and permanent loss of hearing. What many fail to realize, however, is that long-term exposure to even *moderate* sound levels is just as harmful. Bear this mind as you mow the lawn, blast the stereo in your car, or visit a dance club, concert hall, or pub. Think about your ears whenever you swing a hammer, push a plank into a table saw, or drill a hole into a piece of metal.

Foam rubber ear plugs or approved headphones are critical pieces of safety equipment. When selecting hearing protection, look for a decibel value, printed somewhere on the packaging material, that indicates the amount of noise reduction provided. The larger the number, the better.

Protect your lungs! As a young man, I worked for several years at a medical company that leased oxygen equipment. We had two principle customers: ex-smokers, and ex-coal miners. Both had destroyed their lungs inhaling contaminants, and both spent their remaining time on Earth struggling for every miserable breath they took. It's not a fate I'd wish on anyone.

Drilling, sawing, sanding, and grinding operations produce dust. None of this material is very good for your lungs. Masks are cheap, particularly the disposable "surgeon" type. Buy them, and wear them. You'll be surprised at the stains that appear in the filter material, even when breathing supposedly "clean" air.

When using paints, lacquers or other volatile compounds, make sure to work in a well-ventilated area. Lots of solvents are downright corrosive to lung tissue, and ordinary masks won't protect you. The more dilute you keep those fumes, the better off you are. As much as possible, I try to handle those sorts of materials outdoors, or in my garage, with the door wide open.

Protect your hands! Any process that can shape, cut, or drill wood and metal will do far worse to flesh and bone. Be cautious, and exercise good judgement in handling tools and materials. When using knives, cut *away* from your body, not toward yourself. If you're pushing small stock through a table saw, use a piece of scrap wood as a "pusher," not your fingers. Don't hold small parts in your hands when drilling. If the drill bit seizes in the hole, the work will whirl around unexpectedly, causing deep and painful cuts. Avoid wearing jewelry when operating power tools, and avoid loose sleeves or other clothing that could get wound up in rotating machinery. Wear gloves when handling chemicals.

Be alert for fire! Some of the projects that follow involve materials and techniques that, if mismanaged, could result in an accidental fire. Exercise good judgement when using propane torches or similar heat sources. Try to work in a clean area, free of combustible debris. Make sure that soldering irons are unplugged, cool, and properly stowed when you leave your work area. Use care in handling batteries to ensure they are never accidentally short-circuited. Be conscious of the fact that many solvents are extremely flammable. They can be set ablaze with nothing more than an electric spark. Protect your work area with an approved smoke detector (they are cheap!) and a suitable fire extinguisher. *Know how to use it.*

Know what you are working with! In years past, industrial workers often handled potentially hazardous materials in complete ignorance. Fortunately, regulations now mandate that each substance used in the workplace be documented with a so-called "Material Safety

Data Sheet." Material Safety Data Sheets, also known as "MSDS's," describe the physical attributes of various chemicals, as well as crucial information pertaining to storage, handling, disposal, and possible health issues.

Hundreds of thousands of MSDS's are available for free to anyone with access to the Internet. Visit your favorite search engine and conduct a search on the phrase: "Material Safety Data Sheet."

Just for fun, take a look under your kitchen sink and retrieve some laundry bleach or drain cleaner. Write down the chemicals that appear in the ingredients, and then look them up. The exercise is very instructive, if not alarming.

Pay attention! When you're working with tools or conducting experiments, needless diversions or distractions will lead to mistakes. Mistakes may translate into injury. Try to work some place quiet where you won't be interrupted. Unplug the television. Most people watch too much anyway. Don't drink, smoke, or eat anything that's likely to diminish your mental faculties. The most powerful safety device in existence is a clear head and a mind focused on the task at hand.

Don't be afraid to ask for help! I strongly advise younger experimenters to enlist the aid of a mentor. A parent, a teacher, or other responsible adult can provide insight, guidance, and even training. Their wisdom can prevent accidents.

When I think about my own youth and some of the "experiments" I attempted in years past, it makes me laugh, albeit nervously. I don't remember being reckless, it's just that incomplete knowledge and simple inexperience can place a young person at risk. The fact that I reached adulthood unscathed is as powerful an argument for guardian angels as you'll ever find.

While my father was not scientifically inclined, he was an old-school craftsman of the highest order. He went to great lengths to teach me the proper use of tools and some degree of common sense. That alone has probably kept my "bacon out of the fire," and for that I'm eternally grateful.

The bottom line: In this age of blame-shifting and litigation, it's tempting to try to anticipate every possible mistake a reader might make and then issue the appropriate warning. That's a game that no author can win.

Please, use your own common sense. Ultimately, **you** are responsible for what you do, and for your own safety. If something an author suggests does not seem safe to you, **don't do it.** If a process or technique requires more skill than you possess, **don't do it.**

If you are one of those individuals that God blessed with two left hands, please be content to enjoy books like this one for their academic value, and leave well enough alone.

References

http://www.ilpi.com/msds/index.html#Examples
Toreki, Rob, "Where to Find Material Data Safety Sheets on the Internet," copyright 1995-2001, Sponsored by Interactive Learning Paradigms, Incorporated, last revised April 27, 2001

http://www.nsc.org/lrs/statstop.htm
National Safety Council, "What Are the Odds of Dying?" copyright 2000, National Safety Council, last revised July 22, 2000

http://www.osha.gov/
OSHA, "Occupational Safety And Health Administration," copyright ?, OSHA, last revised April 30, 2001

http://www.phys.ksu.edu/area/jrm/Safety/msdshist.html
Kansas State University, "Safety Links: A Brief MSDS History," copyright ?, Kansas State University, last revised November 30, 2000

Chapter 4
What Is An
Amplifier?

What is an amplifier? The question seems silly. Everyone knows, or at least *thinks* they know. In common usage, "amplification" refers to the process of enhancing or enlarging something, but is this definition adequate for our purposes? Before we launch into an analysis, let's consider a few hypothetical cases where amplification is readily apparent.

Case #1: A homeowner, intent on landscaping her front yard, decides that a large boulder must be moved several feet. Throwing her shoulder against it, she mounts an admirable effort, but quickly discovers that there's no way she can induce it to move without assistance. Being a clever and industrious type, she retrieves a length of steel pipe to use as a lever. The pipe is jammed under the boulder, and rested upon a smaller rock that has been placed to act as a fulcrum. This time, when she applies her weight to the free end of the pipe, the boulder yields and rolls in the desired direction.

Case #2: The electrical system in a typical automobile is based upon a 12-volt battery. Current from this battery is delivered to an ignition coil through a switching system that pulses the battery current on and off. Every time a pulse arrives at the ignition coil, the coil responds by producing many *thousand*s of volts. This high-voltage output is eventually directed to the spark plugs, and is responsible for igniting the gasoline vapors during the engine's power stroke.

Case #3: A pleasure boat suffers mechanical problems and stalls in the middle of a large lake. The boat's owner tries to elicit the help of passing boats, but no matter how loudly he shouts, he remains unheard. Finding a newspaper on board, he rolls it into a cone, and

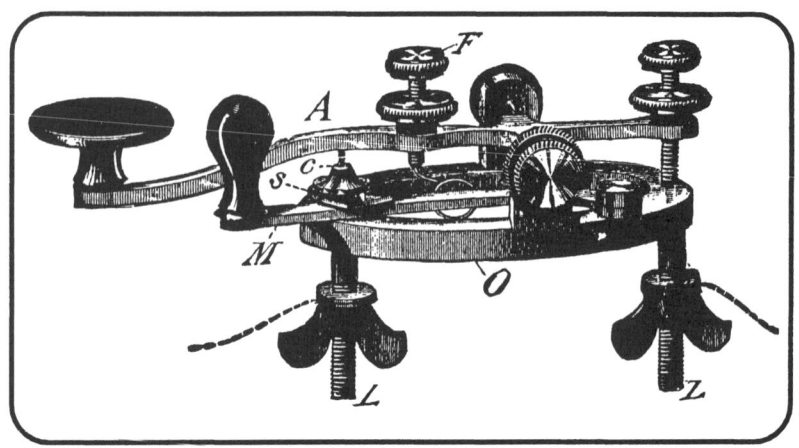

Figure 4.1 A telegraph key
(*American Telegraphy & Encyclopedia of the Telegraph, 1912*)

holds it to his lips. Yelled through this makeshift megaphone, his request for help is suddenly heard at twice the previous distance, and shortly thereafter, a good Samaritan responds by towing the disabled vessel back to shore.

Considering the substance of these stories, I now ask which of them depicts the use of a tool or machine that amplifies? As you'll soon discover, this is a trick question, because the correct answer is, "none of them!" In fact, the cases above provide excellent examples of what amplification is *not*.

Let's review the scene with the lever. It's clear that the short arm of the lever can generate significant force compared to the effort applied to the long arm, but the devil lies in the details. To move the boulder a scant few inches requires the long end of the lever to swing several feet. What's going on here? Well, it turns out that the lever is not so much an amplifier as it is a *conversion* device. It allows you to "purchase" increased force at the cost of displacement (movement).

If you prefer, here's a way to express the argument numerically: Take the length of the long arm of the lever, and multiply that by the force applied. Next, do the same for the short arm (i.e., multiply its length by the weight of the boulder). Compare the two numbers, and you'll discover that they *match*. You get no more out of a lever than you put in to it!

The second case involved an automotive ignition system. An ignition coil is a type of transformer. Inside, there are two coils of wire, both wound on a common steel core. One coil, the primary, consists

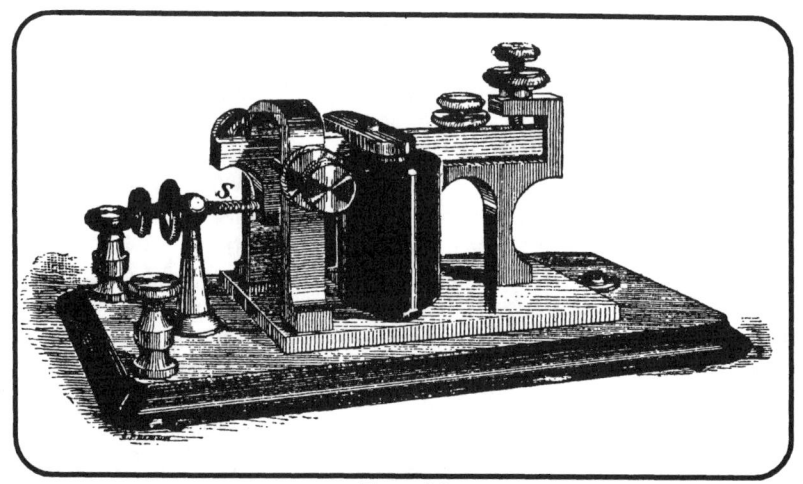

Figure 4.2
A telegraph sounder
(*American Telegraphy & Encyclopedia of the Telegraph, 1912*)

of a comparatively few turns of thick wire. The second coil, or secondary, is comprised of thousands of turns of fairly thin wire. When the primary coil is pulsed by the battery, it generates a field that temporarily magnetizes the core. When the current is interrupted, the field collapses, transferring energy to the secondary coil.

The voltage that appears at the terminals of the secondary coil depends upon the ratio of turns between it and the primary. In other words, if the secondary coil has 1,000 times as many turns as the primary, the battery voltage will be increased by a factor of 1,000. Zap! Twelve volts becomes 12,000 volts! Is there some kind of catch to all this? Of course! What we failed to account for was the magnitude of the *currents* flowing in each coil.

The currents flowing through each coil are affected by the turns ratio just like the voltages are, but in an inverse fashion. The high-voltage secondary can only deliver 1/1,000th the amount of current consumed by the primary.

Confused? Try this: Measure the current drawn by the primary, and multiply that by the battery voltage. That number represents the amount of electrical power being fed into the transformer. Next, multiply the voltage appearing at the secondary by the amount of current (amps) it delivers. This gives you the power going out. If you account for losses in the conversion process, and the variations in voltage and current with respect to time, the computed power values will, once again, match!

Like the lever discussed previously, transformers do not amplify. They are merely conversion devices. They can exchange amperes for volts and volts for amperes, depending upon the ratio of turns between the primary and secondary coils. When all is said and done, you can draw no more power from them than you put in.

What about the third case, the megaphone? If it's not an amplifier, then what is it? Allow me to argue the point this way:

A swimmer wearing rubber fins will easily outpace an equal swimmer with bare feet. Swimming is about transferring mechanical energy from a person's body to the surrounding water. A kicking foot has the tendency to slice through water without properly engaging it. Swim fins, on the other hand, couple more effectively with the surrounding water. They can't make you more powerful, but they do ensure that more of the energy you produce results in useful propulsion.

I consider a swim fin to be a type of "matching" device. It provides an improved interface between a bare foot and the surrounding water, to optimize the transfer of power from one to the other.

This takes us back to the megaphone story. To shout, or for that matter to speak, one must agitate the air to produce sound waves. The mechanism responsible for this is the human larynx. A paper cone cannot, in any sense, amplify a person's voice. Instead it functions, like the swim fins, as a matching device. It improves the interface between one's vocal tract and the free air, thereby optimizing the transfer of energy from the former to the latter.

At this point, you'd probably expect an explicit definition of amplification, or at least a clarification, but I'll defer. With all due respect to Mr. Webster, I'd prefer to take one step back and try to develop the concept from the ground up. To do this, let me introduce one of history's most important inventions, the telegraph.

In its crudest form, the telegraph is composed of three principle components wired together in series, namely, a power source, a telegraph key, and a sounder. For the purpose of our discussion, consider the power source to be a simple battery. When depressed, the telegraph key energizes the circuit, and when released, the current is interrupted. Its function is not unlike a common doorbell button.

The sounder is slightly more complex. It consists of a set of electromagnets, and an iron armature that is suspend by a pivot over the coils. When the coils are energized, the armature is attracted downward until it strikes a metal stop, producing a distinctive "clunk" sound. When the current is interrupted, the electromagnets release the armature, and it returns to its original position by the action of a small spring. Figures 4.1 and 4.2 show both a vintage telegraph key and sounder.

In practice, the sounder is installed at some remote location, and it's linked back to the key and battery with long wires. By pressing and releasing the key in specific patterns, the sounder can be made to clatter and clunk many miles away. A suitably trained listener at the receiving end can easily decode those sounds, extracting the embedded Morse code message.

Nothing in the physical world is perfect. All wire, regardless of its composition, offers some resistance to the flow of electricity. As the distance between a telegraphic sender and receiver grows, and the telegraph wires are lengthened, this resistance increases proportionately. Eventually, a point is reached where the battery at the sending end of the system can no longer overcome the resistance of those lines, and the sounder will fail to actuate. This establishes a practical limit to the distance over which telegraph signals can be sent.

Given all this, let's now consider a hypothetical problem and its implications. In the following scenario, you are the owner of a small telegraph company headquartered in Appleton. You'd like to establish communications between your home base and the town of Zooterville, a span of some 8 miles. While you can easily string wires that distance, electrical resistance limits your equipment's range to no more than 5 miles. What can you do?

One solution is to set up an office in Slackjaw, the little burg located midway between the two cities (figure 4.3). You set up the Appleton sounder there, and hire a telegrapher to monitor it. You also install an additional key and another battery, which is then wired to a sounder placed in Zooterville.

Now, sending a message from Appleton to Zooterville becomes a two-step process. First, the Appleton operator taps out the original message on his key. The signal travels 4 miles to Slackjaw, and actuates the sounder there. The Slackjaw operator listens and decodes the message. Once complete, he retransmits the same Morse code on his own key, which forwards the message 4 more miles to Zooterville.

Strictly speaking, the scheme is a success, but it's not long before difficulties arise. The Slackjaw operator turns out to be a lazy chap who often fails to copy lengthy messages in their entirety. Worse,

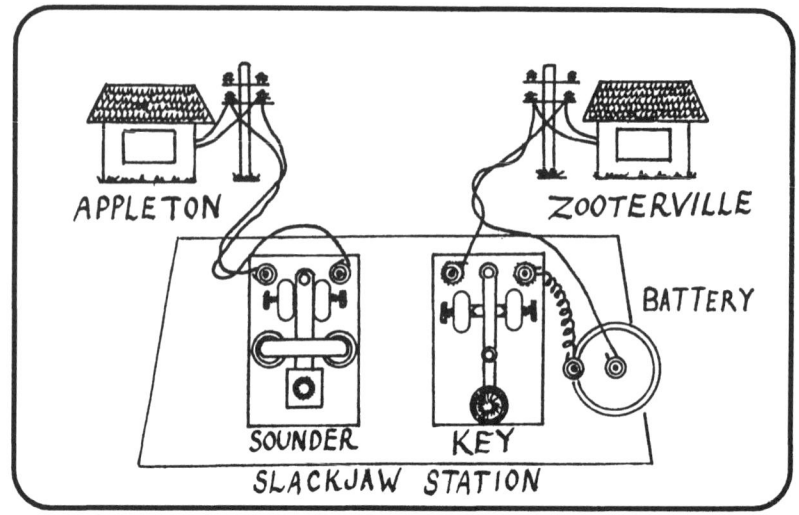

Figure 4.3
The Slackjaw station

he's frequently late for work, if he shows up at all. On several occasions, you've sent important messages from headquarters, only to discover later that they were never echoed to Zooterville. It's time for some drastic changes.

First, you fire the Slackjaw operator. Then, you rig his sounder and his key so that they're mechanically linked together. Now, every time the Appleton sounder clunks, it simultaneously presses the Zooterville key and automatically retransmits the message!

The idea is a brilliant one, and gets even better as the concept is refined. You can modify the Slackjaw sounder by removing the heavy armature altogether. There's no need to keep it, because there's no longer any need for the sounder to make noise. Next, you fashion a tiny switch as delicate and lightweight as possible. The switch is mounted above the electromagnets in place of the original clunky armature. The sounder coils' magnetic field can now act directly upon the switch to open or close its contacts. Operating speed, sensitivity, and reliability are dramatically improved. You've created an important and powerful instrument called a *relay*. Figure 4.4 is an image of a genuine telegraph relay embodying the features I've just discussed. The addition of a relay to your telegraph circuit gives it some interesting properties.

Figure 4.4
A telegraph relay
(*American Telegraphy & Encyclopedia of the Telegraph, 1912*)

Consider the sounder, clunking away at the Zooterville station. Where is the energy coming from that drives its armature? Before answering, review the circuit carefully. While the message certainly originates in Appleton, the current flowing in the coils of the Zooterville sounder actually comes from the batteries housed in *Slackjaw*.

When we began this exercise, we stated unequivocally that our signaling equipment had a 5-mile limit. Suddenly, we're transmitting messages 8 miles with impunity. How? Well, the relay in Slackjaw is able to take the weak signal arriving from Appleton, and use it to control the current supplied by its local batteries. In doing so, it fortifies the signal before sending it on its way.

At last, a definition! An amplifier, in its most elemental form, is nothing more than a machine that allows one source of energy to control another. Typically, the first source is a feeble external signal of some type. The second is a more robust local source of energy. The weak incoming signal is applied to regulate the output of the local supply, the result being a regenerated and enhanced *copy* of the original input. The input has been *amplified!*

At the risk of redundancy, let me restate the critical requirement for amplification. The process always involves two

Page 25

sources of energy, one of which is used to control the other. Bearing this in mind, if we now roll back to the beginning of this chapter, we can reevaluate the opening paragraphs with a fresh eye.

In the case of the lever, there is but one source of energy, the lady landscaper. The lever may transform or convert, but ultimately it's the woman's labor that moves the rock, and nothing more.

The same is true of the ignition coil example. The automobile's battery provides the input, but there is no auxiliary source of energy with which to amplify. In the end, the coil's output is not an amplified copy of the input signal, it's simply an altered form of the original.

A paper cone has no internal power supply. Nothing more can emerge from the large end of the cone than went into the small end. There should be no need to debate the matter of the megaphone.

Back to our fictitious telegraph company. Is there any limit to the distance your messages may now be transmitted? A system using the hypothetical equipment above could wrap around the earth. Admittedly, it would take more than 5,000 intermediate stations, each with relays and batteries. The logistics of keeping that much equipment functional is staggering, and I won't even address the matter of trying to place stations at five mile intervals across an ocean! Those technicalities aside, there's no theoretical reason why an extended line of any length couldn't be installed and used with success.

Let's stretch our imaginations just a bit farther. The obvious usefulness of your relay would probably inspire competitors to design their own. Somebody might even fabricate a device that's markedly better than yours, but how would you ever know? How can one amplifier be compared to another?

If we take the measured output of an amplifying device, and divide that by the measured input, we get a ratio that describes how much amplification is taking place. This numerical value is called *gain,* and it's probably the single most important gauge of an amplifier's usefulness. How much gain does the relay equipment at the Slackjaw station provide?

Assume for the sake of argument that the batteries at the Slackjaw station produce 12 volts. Every time the Slackjaw relay closes, a 12-volt pulse is sent off to Zooterville. The Slackjaw station, therefore, produces a 12-volt output. Suppose, as well, that the incoming signal from Appleton measures a mere 3 volts. This is the weakened input signal that drives the relay.

Divide the output (12 volts) by the input (3 volts) and you get a gain value of 4. This means that the station amplifies incoming messages by a factor of 4 times. Or, stated another way, the signal leaving Slackjaw is *4 times* larger that the signal that came in from Appleton.

The relay is a critically important device that finds application even in modern electronics. You'll probably find no less than a dozen of them under the hood of any car, for example. However, it has one attribute that imposes serious restrictions on its usefulness. Relays are basically digital devices. They can operate in only one of two possible states, that is, either on or off. There is no intermediate state. Thus, they're fine for controlling a telegraph sounder, or the horn of an automobile, but they're basically useless for amplifying continuously varying signals like audio.

Because amplification is most often thought of in the context of analog signals, a simple on-off switching device like a relay may seem an inappropriate example of an amplifier. In defense, the relay meets my criterion for an amplifying device in the sense that it controls one source of energy with another, and exhibits measurable gain. It provides a suitable tool with which to introduce the concepts explored in this chapter. Moreover, the core ideas defined here are easily transferable to the instruments that we'll discuss in the pages to follow.

References

Maver Jr., William, *American Telegraphy & Encyclopedia of the Telegraph,* copyright 1912, Maver Publishing Company, New York, reprint copyright 1997, Lindsay Publications, Bradley, IL, ISBN 1-55918-193-1 (information on telegraph keys, sounders, and relays)

Spenke, Eberhard, *Electronic Semiconductors*, McGraw-Hill Book Company, Inc., New York, copyright 1958, p. 114, (a relay model for amplification)

U.P.C. Book Company, *How to Make Things Electrical*, copyright 1922. U.P.C. Book Company, New York, reprint copyright 1994, Lindsay Publications, Bradley, IL, ISBN 1-55918-149-4, p. 300, (simple construction plans for telegraph sounder)

White, Harvey E., *Modern College Physics*, copyright 1956, D. Van Nostrand Company, Inc., New Jersey

Chapter 5
The Microphonic Relay

In the earliest days of radio communications, so-called "crystal" receivers were the state of the art. A crystal set consists of an antenna and ground, a tuning circuit, a detector, and a set headphones.

Tuning circuits varied, but were typically composed of assorted combinations of coils and capacitors. The values of these components were chosen to establish the tuner's resonant frequency. The antenna might capture and feed any number of stations to the tuner, but a properly designed circuit would accept energy only from the desired channel, while rejecting energy from the unwanted stations.

The detector, which also varied from design to design, generally amounted to nothing more than a diode. Its function was to demodulate the radio signal, i.e., extract the audio information that had been previously encoded in the radio signal.

The extracted audio signal was finally directed into set of high-impedance headphones. Donning these, the listener could hear program material transmitted from miles away. Interested readers should consider perusing my first book, *The Voice of the Crystal,* for additional detail on crystal-radio sets, how to build them, and the radio broadcast process in general.

Crystal sets, by design, are passive devices. They use no batteries or other external sources of power. In fact, all of the energy driving the receiver, including the energy emitted as sound from the headphones, must first have come from a single, feeble source: the antenna! The crystal receiver can deliver no more power, audio or otherwise, than can be collected from the "air." The weaker the incoming signal, the weaker the sounds in the headphones.

Intuitively, we know that as we increase a receiver's distance from a transmitter, the signal is weakened. This is due in part to a relationship called the *inverse square law*. Essentially, this rule says that as you increase your distance from an electromagnetic signal source, the signal weakens as the *square* of that distance. If, for example, you double your distance from a transmitter, the apparent strength of its signal will fall to one-fourth its previous value. Double that distance again, and the signal degrades to one-sixteenth.

It quickly becomes evident that even modest increases in the span between a radio sender and receiver results in significant reductions in the strength of the transmitter's signal. This can impose crippling restrictions on the usefulness of a passive receiver. Crystal-radios are remarkable instruments, but as passive devices, their function is entirely dependent upon the strength of the incoming radio waves. Large distances mean weakened signals, which ultimately translates to poor receiver performance. The distance at which desired signals can no longer be heard represents the useful range of the communications link.

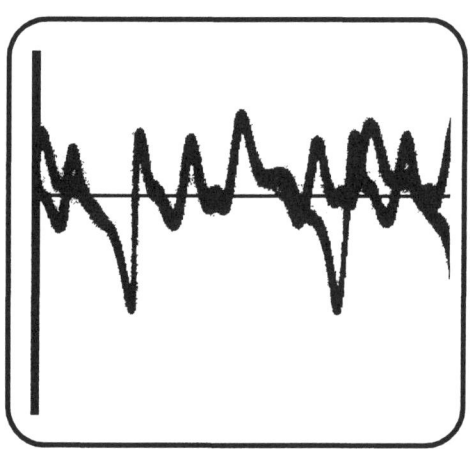

Figure 5.1
The "Ahhhh..." sound

An obvious improvement to the basic crystal set is an audio amplifier. By boosting the audio signal before feeding it to the headphones, it becomes possible to recover faint whispers that might not have been perceivable before. The ability to rejuvenate an incoming program also gives the receiver extra range. It can be moved that much farther from the transmitter before the signal again degenerates to a useless level.

Given the relative success of our telegraphic experiments in the last chapter, it seems reasonable to reach, once more, for the venerable relay. Regrettably, it's not a suitable amplifying device for this type of application. Its downfall lies in its constitution.

A set of switch contacts, when closed, offers very low resistance to the flow of electric current. The measured resistance across it may be some tiny fraction of an ohm. When the contacts open, the resistance soars to many hundreds of millions of ohms, or higher. Switch contacts are generally thought of in terms of "on" and "off," but in a way, they represent a very specialized type of "adjustable" resistor. This resistor has but two settings: negligibly low, or absurdly high resistance.

When used as part of a primitive amplifying instrument like a relay, the all-or-nothing nature of these contacts means that the local power supply, used to "refresh" the incoming signal, can likewise assume only the fully on or fully off status. Thus, the instrument, for all its utility, is relegated solely to digital applications.

By nature, analog signals are much more complex than discrete bursts of on and off. For one thing, they're continuously variable. They may occupy a vast spectrum of voltages, from low to high, and all levels in between. In the case audio

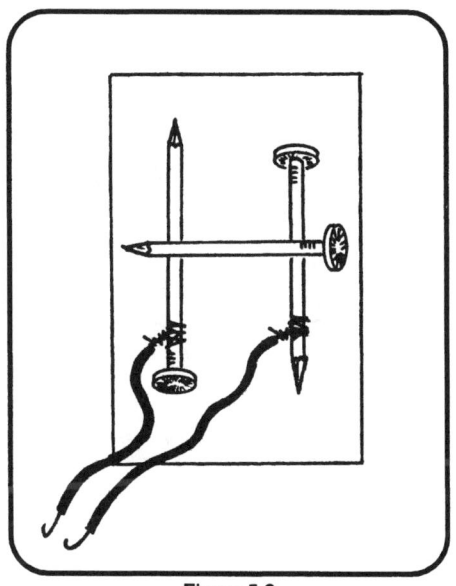

Figure 5.2
An unstable contact microphone

or radio waves, these signals may take the form of simple sine waves or extremely elaborate, repetitive patterns. The complex pattern shown in figure 5.1 depicts the simple vowel sound, "ahhhh," as captured on the screen of an oscilloscope.

If it were somehow possible to engineer a set of contacts, such that their resistance could be varied from high to low, *or any point in between*, it might be possible to build a variant of our relay that could amplify analog signals as well! Is such a thing possible?

In the latter half of the 1800's, a fellow by the name of David Edward Hughes began studying the nature of imperfect contacts between current-carrying materials. In one experiment, Hughes set up an instrument composed of three ordinary nails. Two were laid parallel to one another. The third was placed, like a bridge, across the other two. See figure 5.2. Battery current was passed through the nails, and through what amounted to a headphone or speaker. This configuration proved to be extraordinarily sensitive to vibration. Any agitation of the nail bridge resulted in varied crackling sounds in the headphones. Motion in the nails disturbed the tiny regions of contact between them, causing the current to fluctuate in response to the vibration.

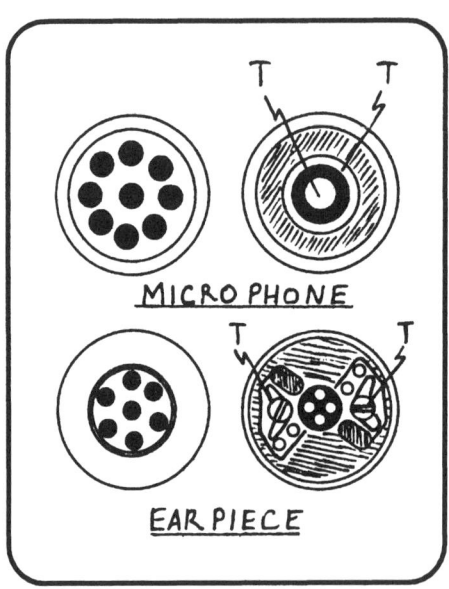

Figure 5.3
Traditional telephone handset components

Hughes showed that his instrument could be made to respond to the vibration imposed by speech or other sound waves. He realized that his unstable contacts did not simply switch current on and off. They were actually capable of *varying* the electric current through them in response to the incoming sound waves, particularly when they were composed of carbon. On the basis of these discoveries, Hughes went on to invent and develop the carbon microphone.

Carbon contacts work well in this application, because the resistance between them varies with the amount of pressure applied to them. When the contact pressure is comparatively high, resistance drops, and currents flow easily. When the pressure is low, resistance rises, so the current begins to drop off. Rising and falling waves of pressure, like sound waves, cause an electrical current flowing through those contacts to rise and fall in lock-step.

In 1876, Thomas Edison refined Hughes' idea. He replaced the single set of carbon contacts with a capsule partially filled with crushed carbon. The capsule was fitted with electrodes at each end to conduct current through the powder, and it was mechanically arranged so that speech vibrations striking a diaphragm could be made to agitate the carbon granules.

The point at which each granule touches its neighbor forms an unstable contact capable of controlling electricity. Together, the carbon particles create a cascade of hundreds of such contacts, each capable of regulating the flow of current. As a result, Edison's design was much more sensitive, and somewhat more tolerant of rough handling than Hughes' instrument.

Traditionally, Bell is given credit for the invention of the telephone, though his instrument would never have been practical but for the invention of the carbon microphone. During most of the last century, every telephone at every desk, kitchen wall, and phone booth contained in its handset a carbon microphone of the type described above.

In more recent years, telephones and their internal electronics have been revised. Modern handsets now rely on newer technologies. The old "standard" phones have become passe, and

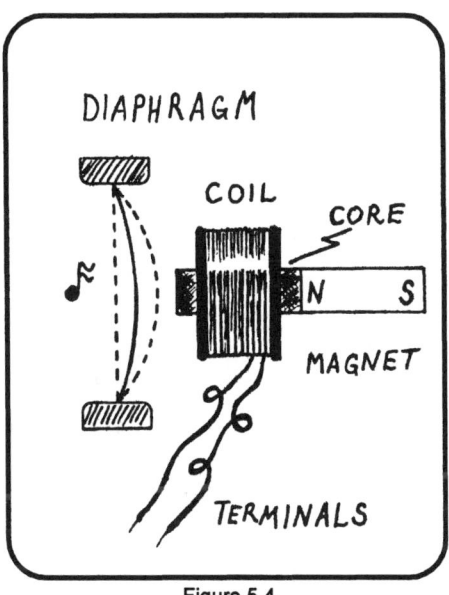

Figure 5.4
Diagram of the operating principle behind earphones

therefore appear with some regularity at flea markets, garage sales, and thrift shops. They can be had for a couple of dollars, and are an excellent source of experimental odds and ends. In fact, it's possible to build a very simple carbon audio amplifier with nothing but the parts of the handset. In the 1920's an instrument of this type was referred to as a "sound intensifier" or "microphonic relay."

Figure 5.3 shows a carbon microphone, removed from the mouthpiece of the phone by unscrewing the cap. The front of the mike is ventilated, so that sound waves may stimulate the diaphragm within.

Inside of the mike are two electrodes, a fixed one made of carbon, and a gold-plated electrode attached to the mike's diaphragm. Between the two is a region filled with granulated carbon.

The back side features a circular metal button surrounded by a metallic ring. The button is one terminal of the mike, the ring constitutes the other. These terminals are marked "T" in the illustration.

For the experimenter, connecting wires to the microphone can be problematic, though there are several options. One is to simply tape wires to the ring and button using electrical tape. This works for a quick test, but tends to be unreliable in the long run. Another approach is to solder wires to the terminals. If attempted, this must be done quickly and with finesse, as excessive heat may damage or degrade the device.

Yet another option is to take advantage of the mike holder found in the telephone handset itself. It's a cup-like affair, featuring two spring "fingers" designed to press against the microphone's terminals. At the rear of the holder are two screws to which wires may be attached. In normal use, the cap on the handset's mouthpiece applies the necessary pressure to keep the

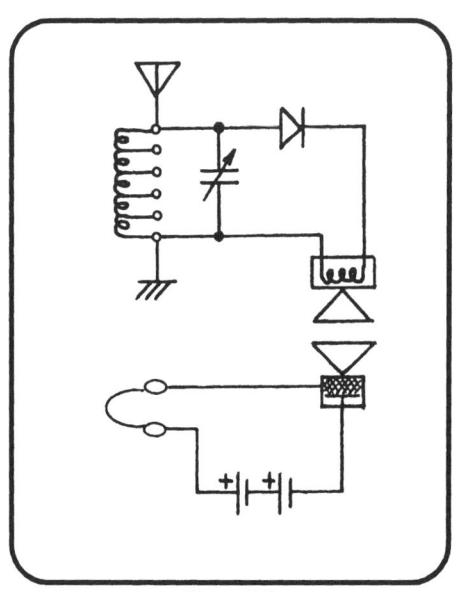

Figure 5.5
Simple crystal receiver with primitive carbon amplifier

microphone seated in its holder. Removed from the handset, a strip of black electrical tape will do the job.

If the handset's earpiece is now unscrewed, you can remove the *receiver*, or *headphone*, also shown in figure 5.3. Notice the screw terminals on its reverse side. The principle behind the operation of the receiver is very similar to that of a common loudspeaker, so from here on out, I'll refer to the receiver as such. For those unfamiliar with the mechanism behind the receiver, let's digress just long enough to touch upon the basic principles.

The specific construction of the handset's speaker depends in part on its make, model, and year of manufacture. All of them feature some form of steel core, upon which is wound a fine coil of copper wire. Near the core is a thin, flexible, magnetic diaphragm. Most models also contain a permanent magnet, whose field permeates both the coil's core and the diaphragm itself.

If an electrical audio signal is fed to the coil, it energizes and creates a magnetic field. The coil's field varies in response to the input signal, which acts to alternately reinforce and counteract the static field of the permanent magnet. The sum of these two results in a composite field that induces the diaphragm to vibrate. The reciprocating motion of the diaphragm produces sound. Figure 5.4 illustrates this principle.

To construct a microphonic relay, now that we have a microphone in one hand and speaker in the other, we simply bring the two together, face-to-face. That's it! For experimental purposes, the assembly can be held together through the creative application of electrical tape. I chose to implement a more permanent arrangement, which I'll discuss in greater detail in just a moment. In any event, it's important to ensure a good, airtight seal between the speaker and the mike.

The "input" to our new amplifier, that is to say, the signal we wish to amplify, is connected to the terminals of the speaker. If that signal is of sufficient strength, the speaker's diaphragm will move, producing sound. Because the speaker and mike are taped together, the sound waves don't have to travel very far before they strike the diaphragm in the microphone. The motion of the mike's diaphragm agitates the carbon granules inside, which causes their electrical resistance to rise and fall in sympathy with the sound waves.

The microphone is connected to a local source of energy, in this case some flashlight batteries, and then to a set of headphones. As the resistance in the microphone varies, it controls the flow of current into the headphones. The original signal is thus reconstructed.

Figure 5.5 shows a sample schematic for a simple crystal receiver. The circuit features a primitive tuner and a 1N34 germanium diode used as a detector. The tuning coil is composed of 50 turns of enameled wire, wound on a 3-1/4-inch cardboard mailing tube. The turns are spaced to occupy about an inch and a half in width. Every tenth turn is tapped to facilitate tuning. The tuning capacitor is of the variable type, selected for a value of 360 pF with the plates fully meshed.

The output of the detector is fed to the carbon amplifier, which is intended to boost the audio signal. Two flashlight batteries power the amplifier, and the headphones can be pretty much any type that you desire, including low-impedance stereo headsets. Using a 15-foot length of wire as an antenna and a decent ground, I was able to tune in several stations with a surprising degree of volume. The amplifier was more than capable of generating useable volume in the cheap set of headphones I happened to salvaged from a discarded cassette player.

Just for fun, let's compare our "microphonic" amplifier with the telegraph relay from the previous chapter to see how the two stack up. We'll start with the input. In the case of the telegraph relay, the weak input signal is fed into the relay coils. A magnetic field results, which attracts and causes the relay's armature to move. In the carbon amplifier, the input signal is fed to the voice coil of the speaker. There, the resulting magnetic field causes the speaker's diaphragm to move, generating a sound wave.

In the telegraph relay, the movement of the armature causes a set of contacts to close and open. In the carbon amplifier, the sound waves from the speaker strike the microphone's diaphragm, agitating the carbon particles inside. As pressure waves travel through the carbon granules, their collective resistance rises and falls with them.

When the relay contacts close and open in response to the incoming message, they activate and deactivate the local supply, reconstructing and regenerating the original Morse-code message. (The relay echoes the incoming patterns of "on" and "off.") In the case of the carbon amplifier, resistance changes in the carbon particles allow them to regulate the local supply to high, low, and an infinite number of intermediate values. In doing so, the local supply is controlled in such a fashion that its output mimics the original signal.

Building A Microphonic Relay

The mechanics behind the Microphonic Relay, when constructed of telephone parts, is so simple that there's no reason not to build a permanent version of the device. In fact, once I had developed a formula, I rapidly produced four of the instruments depicted in figure 5.6.

In each instance, the base of the instrument was fashioned from a round, decorative plaque, 5 inches in diameter. The plaques were purchased in the raw from a local craft store. After they had been modified as described below, the wood was stained and sealed with polyurethane.

Figure 5.6
A homebrew "Amplifying Relay"

The base was drilled and countersunk to clear two 6-32 tee nuts, which were pressed in from the bottom. Threaded into the tee nuts were two segments of 6-32 threaded rod. The rods were spaced about 2-1/2 inches apart, and were cut to project vertically from the base to a height of 4 inches. Each rod was secured with a washer and nut, tightened onto the top surface of the base.

A steel bridge was fashioned from 1/4-inch steel bar stock. The stock was cut to length of 3 inches. I bored three 1/8-inch holes through the bridge, one through the exact center, the others spaced 1-1/4 inches to either side of it. The base, vertical threaded rods, and the bridge form a clamping structure designed to hold the stack of telephone components in proper alignment. In the final assembly, 6-32 wing nuts were used to compress the stack.

The body of the instrument is composed of two 1-1/2 inch copper pipe caps. Each cap was bored twice, a 1/8-inch hole in the center of the face of the cap (for mounting purposes), and a 1/4-inch hole drilled through the side (to allow the passage of wires).

One of the caps was placed on the base of the instrument, open side up, centered between the vertical rods. A 1/2-inch, #6 sheet-metal screw was used to anchor the cap to the base. A mike holder, salvaged from a telephone headset with the mike, will slide into the pipe cap with little effort. All that's required for a perfect fit is the removal of a small protrusion from the side of the holder. A sharp pocketknife or razor blade can do the job nicely. Once proper clearance and fit was assured, the holder was removed so that lead wires could be connected to its terminals. The wires were routed through the wall of the cap via the 1/4-inch hole. The holder was then pressed into place.

Next, the actual carbon microphone was set into the holder. I purchased a 1-1/2-inch neoprene "O"-ring, and located it on the face of the mike. When the speaker is placed face down onto the growing stack, the "O"-ring provides an excellent seal between the two elements, and offers some cushioning as well. Lead wires were attached to the speaker terminals and left dangling.

The second pipe cap was bolted to the bridge with a 6-32 machine screw and a wing nut. The bridge and cap assembly was then lowered onto the rear face of the speaker. The speaker leads were routed through the 1/4-inch hole in the wall of the cap.

Wing nuts were installed on the threaded rods, and tightened just enough to hold the various components together. The microphone, speaker, "O"-ring, and caps were tweaked and adjusted for proper alignment. The wing nuts were tightened an extra turn or two to lock everything into place. Figure 5.7 is an exploded view of the instrument showing the relationship between its primary parts.

Finally, the microphone and speaker leads were terminated with Fahnstock clips, secured to the base with 1/2-inch #6 sheet-metal screws. I like the look of vintage cotton-covered wire, which is now

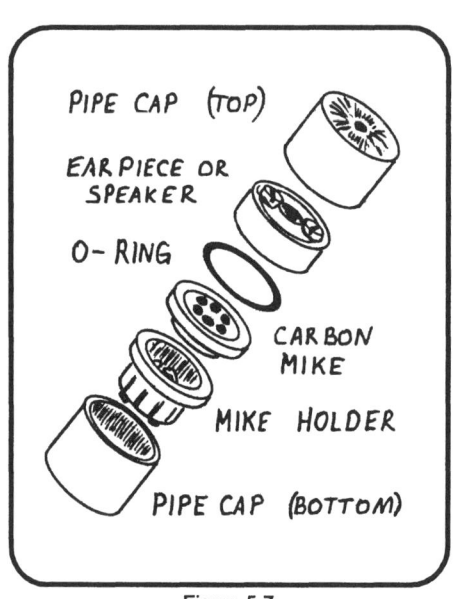

Figure 5.7
Exploded view of the amplifying relay's internals

PIPE CAP (TOP)
EAR PIECE OR SPEAKER
O- RING
CARBON MIKE
MIKE HOLDER
PIPE CAP (BOTTOM)

difficult to get. I made my own using the "shoestring" method described in *The Voice of the Crystal.*

End Results and Going Further

The carbon microphone amplifier discussed here actually works, though it has some limitations worth discussing.

The first concern lies with the speaker. Crystal-type receiving circuits generally work best while driving headphones with as high an input impedance as possible. High-impedance headphones draw energy from the tuner sparingly and produce useable amounts of sound with a minuscule amount of electricity. Speakers salvaged from telephone headsets, on the other hand, were never intended for this type of service, so they are typically designed with a much lower impedance. While they will function in the circuit described above, they also have the tendency to "load" the tuning circuit, interfering with its ability to sharply select desired stations. In other words, you may end up with more than one station playing at a time, regardless of how your tuning components are set up.

There are several ways to address this problem. One option is to drive the amplifier from one of the tuning coil's taps. By using a tap in the coil as shown in figure 5.8, you can reduce the extent to which the carbon amplifier burdens the tuning circuit. Regrettably, the audio volume may also be reduced.

A second option is to replace the telephone speaker with something more efficient. If you can find or salvage some other type of speaker with a greater impedance, you can use it to your advantage. From time to time, I've come across old high impedance headsets in which one of the earpieces was missing or destroyed. The remaining earpiece could easily be recycled for use in the carbon amplifier, and it would probably improve its operation. Brand-new high-impedance headphones are still available from a number of suppliers, and are inexpensive enough that you might entertain the prospect of buying a set, just to harvest one of the earpieces.

Another concern with this amplifier design is the carbon microphone element. Not all elements are the same. There seems to be some variation in their sensitivity and electrical characteristics, depending upon which year they were manufactured. Don't settle on the first one you come across. I've also noticed some variation in sensitivity based on the orientation of the mike (vertical vs. horizontal). All of the mikes I've played with show the greatest sensitivity when they're oriented face up, though they are also most unstable in this configuration. Face down, they seem to exhibit the most robust tone, though sensitivity is markedly reduced. A good mike that suddenly becomes unresponsive may be restored with the sharp rap of a pencil. I suspect that the carbon grains may, under some circumstances, settle or become packed and matted together. Minor mechanical shock seems to set them free.

It has occurred to me that an interesting modification to my amplifier design would be to set it up on a pivoted frame or gimbal, as is sometimes done with hourglasses. The microphone/speaker assembly could then be set upright, inverted, or set to any intermediate angle desired. Thus, the tone and sensitivity attributes of the instrument could be tailored to the specific application.

In the end, carbon telephone mikes were designed to respond to human speech at the levels of normal conversation. There appears to be an audio threshold below which they don't work very well, so you may discover that extremely weak radio stations are simply not processed by this equipment at all.

One interesting effect you may notice is a muted hiss, heard in the headphones whenever the amplifier is in operation. It does not come from the input signal. In fact, it will likely be present even with the input completely disconnected.

All amplifiers suffer from the problem of internally generated noise. In this case, the hiss is literally the sound of electricity passing through the carbon granules in the microphone. Internal noise is a bad thing, because it waters down and degenerates the desired signals. When amplifiers are cascaded, that is, one chained to another for greater amplification, the situation worsens. Any internally generated noise gets amplified in later stages, along with the intended signals. Eventually, the noise may rise to a point where you can no longer distinguish between it and the legitimate signals you intended to amplify.

The better an amplifier is, the less internal noise it generates. Amplifier quality in this regard is measured by its *signal-to-noise ratio.* The signal-to-noise ratio is calculated by dividing the amplitude of the desired signals by the level of the internally generated noise. The higher the numerical result, the quieter the amplifier is in operation, and the better the amp.

You can, to some extent, vary the amount of amplification by toying with the local power supply. I recommend something on the order of 1.5 to 3 volts (one or two flashlight batteries in series), but you may find higher voltages more effective. I have used these instruments with 4.5 volts, but I wouldn't go any higher. Excessive currents cause the carbon elements to become noticeably *less* sensitive, probably due to local heating effects. With it, there is a noticeable increase in noise, and a real danger of permanent damage to the microphone elements.

I alluded to the fact that I built multiple microphonic relays, four to be exact. I did so, to experiment with an amplifier chain. The tuner circuit described earlier was wired to the first amp, which in turn drove the second amp. The second drove the third, the third drove the fourth, and the fourth powered a set of headphones. As expected, the total amplification through the system was, under some circumstances, quite significant, but some interesting pitfalls were discovered.

While more stages provide more gain, the effort yields diminishing returns after the addition of the third amplifier. The cascade of amplifiers, as expected, generated a greater amount of hiss than a single stage. Finally, I concluded that the system worked best with separate power supplies (batteries) for each stage of amplification. Running all the stages from the same set of batteries allowed for a peculiar interaction between the various stages. This resulted in the production of unexpected squeaks, whistles, and howling in the headphones.

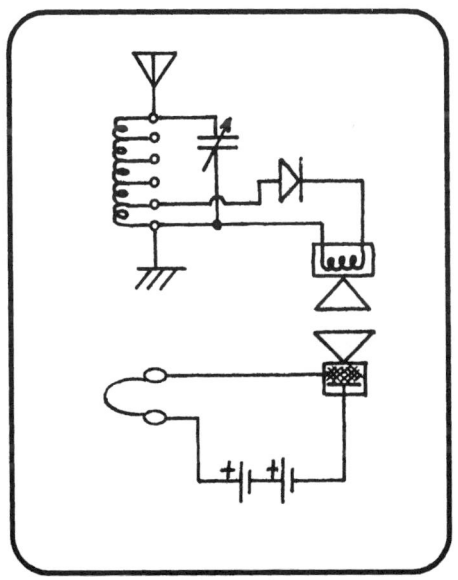

Figure 5.8
Reducing the load on the tuner circuit

References

Adams, Joseph H., *Harper's Electricity Book for Boys,* copyright 1907, Harper and Brothers Publishers, reprint copyright 1998, Lindsay Publications, Bradley, IL, ISBN 1-55918-206-7, pp.168-177, (construction details for two different carbon-granule telephone transmitters/microphones)

Diers, Tracy, "Meet Mr. Mike," from *How To Build 78 Radio and Television Sets,* copyright 1950, Popular Science Publishing Co., Inc., pp. 46-50, (construction details for various types of microphones)

Friedrichs, H. Peter, *The Voice of the Crystal - How To Build Working Receiver Components Entirely From Scratch,* copyright 1999, H. P. Friedrichs, ISBN Number 0-9671905-0-9, (basic radio theory, extensive information for crystal-radio projects)

Hawkins and Staff, ed., *Hawkins Electrical Guide Volume 7,* copyright 1917, Theo Audel & Co., New York, pp. 2,117-2,119, (details of carbon button telephone transmitter/microphone)

Hopkins, George M., *Experimental Science, Volume II,* copyright 1902, Munn & Co, New York, reprint copyright 1997, Lindsay Publications, Bradley, IL, ISBN 0-917914-50-3, pp. 61-67, (discussion of the Blake carbon-button telephone transmitter, and two experimental unstable contact microphones)

Morgan, Alfred, *The Boy's First Book of Radio and Electronics,* copyright 1954, Charles Scribner's Sons, New York, pp. 202-205, (experiment with unstable carbon contact microphone)

U.P.C. Book Company, *How to Make Things Electrical,* copyright 1922. U.P.C. Book Company, New York, reprint copyright 1994, Lindsay Publications, Bradley, IL, ISBN 1-55918-149-4, p. 300, (super sensitive microphone based on unstable carbon contacts)

Windsor, H. H., *The Boy Mechanic - 700 Things for Boys to Do,* copyright 1913, Popular Mechanics Press, Chicago, reprint copyright 1988, Lindsay Publications, Bradley, IL, ISBN 0-917914-88-0, p. 398, (construction details for a homemade carbon-granule telephone transmitter/microphone)

Chapter 6
The Balance-Beam
Amplifier

Considering their relative simplicity, carbon-contact amplifying devices worked pretty well. According to texts from the period, several amplifying devices could be connected together in a cascade or chain, such that each instrument amplified the output of its predecessor. In this configuration, weak audio signals could be boosted to 20 or even 30 times their original strength. Transoceanic communication, using amplifiers of this type, was reported to be quite good. Conceptually, these devices differ little from the microphonic relay discussed in the last chapter. The key to their performance was a matter of streamlining and optimization.

Figure 6.1 shows a wiring diagram for the Brown Amplifying Relay, one of the more successful microphonic relays. Note that the instrument contains neither a "speaker" nor a "microphone" as such. Rather, both have been combined and fused together to form a single, efficient instrument. An image of the instrument appears in Figure 6.2, as seen in Bucher's *Practical Wireless Telegraphy*.

The input signal, applied to terminals "A" and "C," energizes a set of coils labeled "W1." Instead of a speaker diaphragm or iron armature, as was the case in the telegraph relay, the magnetic field is applied to a delicate metal reed. At the tip the reed is a tiny button of carbon. The carbon rests against a fixed contact "C," composed of a special osmium-iridium alloy. The osmium-iridium and carbon form a classic unstable contact microphone, that responds to vibrations set up in the reed. Ironically, this architecture bears a striking resemblance to the basic telegraphy relay, with the significant difference that the Brown Amplifying Relay is capable of processing *audio* signals.

It appears to me, as the result of my own experiments, that the pressure range over which a set of carbon contacts will faithfully reproduce signals is both small and well bounded. Excessive pressure yields poor sensitivity, while insufficient force results in disagreeable noises and unpredictable behavior. Optimum performance occurs at a precarious point bordered by rather narrow limits.

By all accounts, carbon contacts are notoriously difficult to adjust properly, and they represent a severe challenge to maintain in a working condition. The combination of osmium-iridium and carbon contacts in the Brown device may have been an attempt to widen the useable range of operating pressures through the application of exotic materials.

You may also have noted the extra set of coils, marked "W2." My interpretation of the schematic suggests that if the pressure between the contacts was excessive, the current in "W2" would rise, tending to draw the reed away, and relieve that pressure. In other words, I suspect that the additional coils and circuitry were part of an attempt to make the instrument self-adjusting.

In this chapter, I'd like to introduce my version of the Brown Relay, which I call the "Balance-Beam Amplifier." This instrument is reasonably simple to build, being composed of common materials. Nonetheless, it is extremely sensitive, relying on the action of gravity and a delicate balance to establish the proper operating pressure between its carbon contacts. Figure 6.3 shows the completed instrument. It's composed of four major subassemblies: The Driver Assembly, The Diaphragm / Lower Contact Assembly, the Balance-Beam / Upper Contact Assembly, and the Base and Terminal Assembly. Let's dive right in with a discussion of these components, their respective functions, and how to fabricate them.

Driver Assembly

The driver assembly is composed of a set of wound copper coils, two magnetic pole pieces, two permanent magnets, and a steel back strap. Its purpose is to convert the amplifier's input signal into a varying magnetic field that can be used by other parts of the amp to manipulate a set of carbon contacts.

Like any of the typical microphonic relays, the Balance-Beam Amplifier requires a set of electromagnets to operate. I have manufactured a variety of electromagnets for other projects, and generally begin by fashioning some type of spool on which to wind the wire. This time, I had hoped to find something suitable that could be used as-is, and found the solution in a local fabric store.

The spools on which common sewing thread is sold are relatively large, and contain generous quantities of thread. For use in electric sewing machines, the thread is frequently transferred from the large spools to smaller bobbins. Empty bobbins are sold for this very purpose. They're available in standard sizes, depending upon the manufacturer, and are produced both in steel and in plastic. The spools I selected were clear plastic, manufactured by a well known sewing machine maker. The bobbins have an outside diameter of 13/16 inch, a depth of 7/16 inch, and feature a spindle hole that is 1/4 inch in diameter. You'll need four of them. See figure 6.4.

My coils were wound with a fine enameled magnet wire, salvaged from the coils of a junked telephone ringer. You can harvest similar wire from a scrapped phone, or a discarded electric guitar pickup. (If you choose the latter, make certain that you are not destroying a collectable or valuable antique! See my tips and warnings in *The Voice of the Crystal*.) I measured the diameter of my wire with a micrometer, and estimate it to be about #40 gauge.

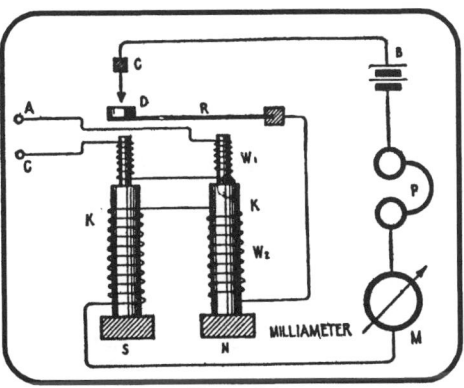

Figure 6.1
Brown's amplifying relay - circuitry

To wind the coils, I slid a 1/4-inch bolt through each bobbin, fitted a nut to it, and then chucked the bolt in my rechargeable electric hand drill. The supply spool was slid onto the shaft of a small screwdriver, which in turn was clamped in a vise. This allowed the supply spool to spin freely as wire was withdrawn from it.

As I mentioned in an earlier chapter, my hand drill has an exceptional trigger speed control, which allows me to carefully regulate the accumulation of wire on the bobbins. The implied goal is to wind the coils rapidly. However, thin wire is extremely prone to breakage, which limits your speed. Thin wire also has the tendency to whip itself into tiny kinks and knots which must be avoided at all costs. Exercise extreme caution when drawing thin wire between your fingers, because it can slice into flesh like a vicious little bandsaw.

I made no attempt to count the number of turns wound on each bobbin. Rather, I accumulated wire in as even and compact a fashion as possible. When the layers of wire had risen to within 1/16 inch of the edge of the flanges, I judged the bobbins to be full. Using a short piece of fabric adhesive tape, I wrapped the windings once to stabilize and protect them. For the moment, the free ends of the wire were left dangling.

Forty-gauge wire is far too delicate to leave exposed. Repeated flexing or tension on wire leads can easily break them off. To prevent this, the leads must be terminated with something more robust. I used thin, plastic insulated, stranded wire. The wire was removed from some scrap electronics, and is probably #28 gauge or similar. You can sometimes find nice, light-gauge stranded wire in the tone arms of old record players. Or, you can try cutting apart the cable from a discarded computer mouse.

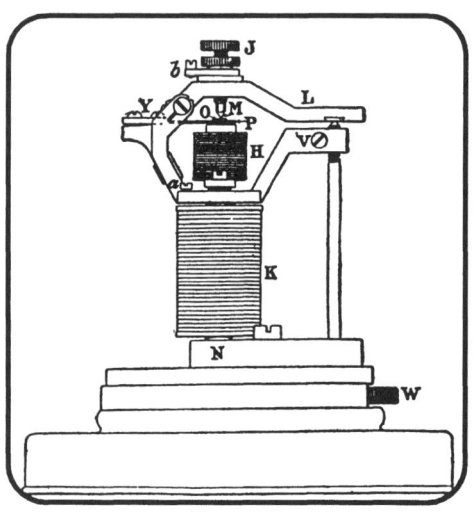

Figure 6.2
Brown's amplifying relay - a side view

Figure 6.5 shows how to terminate a coil. First, two stranded leads are stripped, tinned with a soldering iron, and bent into the shape of little hooks. The hooks are held in place with a piece of tape. The coil leads are soldered to the hooks, and then the entire coil body is then wrapped tightly with tape. Because polarity will become important later on, use *two different* colors for the stranded wire. Each of my coils has a black lead, and a white lead. (At this point it doesn't matter which color is wired to which end of the electromagnet, so don't worry about it.)

When you're done, check each coil with an ohmmeter. My coils measured about 600 ohms apiece, and yours should be in this ballpark. An extremely low reading means you have a short circuit, or

Figure 6.3
The Balance Beam Amplifier

used wire that's too heavy. An infinite reading means the wire is broken someplace. In either case, you'd better unwrap the tape, unwind the coil, and start over.

Each coil must now be tested and marked for polarity. This is done with an "AA" battery, a 1/4-inch bolt, and a magnetic compass. Take the bobbin, and slide a bolt though its center. The bolt will act as a temporary core. Connect the black lead of your coil to the negative terminal of the battery, the white lead to the positive. Now, bring the compass near both ends of the bolt, and observe the movement of the needle. You must determine which side of the coil attracts the north-seeking point of the compass needle. Once you've identified the correct side, you need to mark it. I marked my coils with a tiny blue dot from a paint pen. See figure 6.6. Repeat the process for each of the four coils. This completes the set, and they can be put aside.

The pole pieces are made from 1/2-inch steel rod, commonly found at hardware stores. Instead of steel rod, 1/2-inch bolts would probably work just fine. The first step in fabricating a pole piece is to cut off an appropriate length. In my prototype, this amounts to an inch and a half.

EMPTY SPOOL

WOUND

Figure 6.4
Coil bobbins

Figure 6.5
Attaching leads to the coil

Next, the rod must be machined to the shape depicted in figure 6.7. Note that about a third of the pole piece is left its original diameter. The rest is turned down to a diameter small enough to slide into the plastic coil bobbins. The exact diameter depends on the brand of spool you select, but it will be on the order of 1/4 inch.

Part of the machining process involves "facing" both ends of the pole piece. This means that the ends of the rod should be flat, smooth, and as much as possible, perpendicular to the axis of the rod. Finally, the thick end of each pole piece must be drilled and then tapped to accept a 6-32 machine screw.

The easiest and most accurate way to accomplish the steps above is to use a small lathe. The second best approach is to chuck the rod in a drill press, and carefully work it with a metal file as it rotates. You won't get the precision and uniformity that a lathe delivers, but you'll probably get close enough.

If you don't have a drill press, the next best bet is to try to use an electric hand drill. The drill motor is clamped in a bench vise, and the rod is loaded into the drill's chuck. As is the case with the drill press, the rod can be worked with a metal file as it spins.

If all else fails, use a 1/4-inch machine bolt as the pole piece, (use the kind with the hexagonal head). The head of the bolt can be drilled and tapped for the 6-32 mounting screw.

Trial and error suggests that better pole pieces result if they are annealed. To anneal the pole pieces, they should be suspended in the flame of a propane torch and heated until they glow. Next, they must be removed from the flame and placed on a nonflammable surface to cool. An old brick, a ceramic dish filled with dry, clean sand, or similar material will work just fine. The cooling process should be as gradual as possible. Don't quench the pole pieces in water or fan them in an attempt to cool them.

The driver assembly requires two permanent magnets. The magnets I selected are extremely common, and can be harvested from the magnetic catch hardware found in many kitchen cabinets. Any home-improvement store will carry magnetic catches. Some stores actually carry the naked magnets themselves.

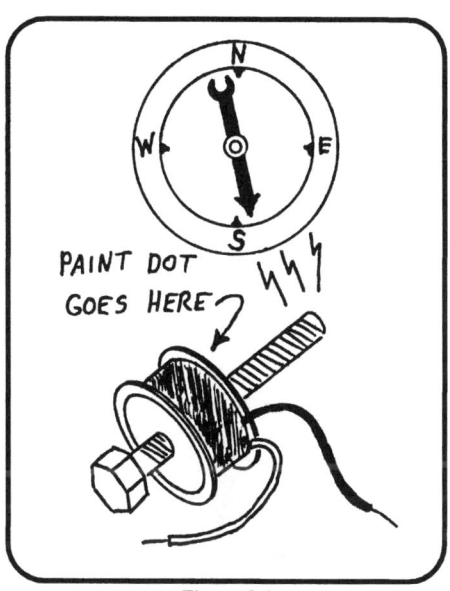

Figure 6.6
Determining coil polarity

Catch magnets are flat, rectangular, and made from a ceramic material. Typically, they're pierced through the center with a mounting hole. The north and south poles of the magnet are found on the large faces. See figure 6.8. To use these magnets, it's necessary to identify their poles. Like the coils, this can be done with a compass. Simply determine which face of the magnet attracts the north-seeking point of the compass needle and mark it. Again, I tagged my magnets with a small blue paint dot. Be careful. It's possible to ruin a compass by

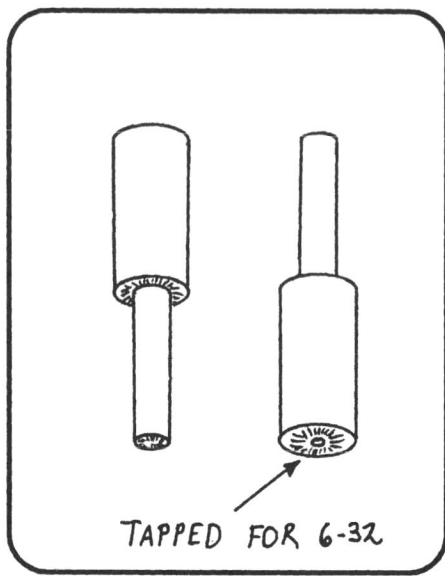

TAPPED FOR 6-32

Figure 6.7
Pole pieces

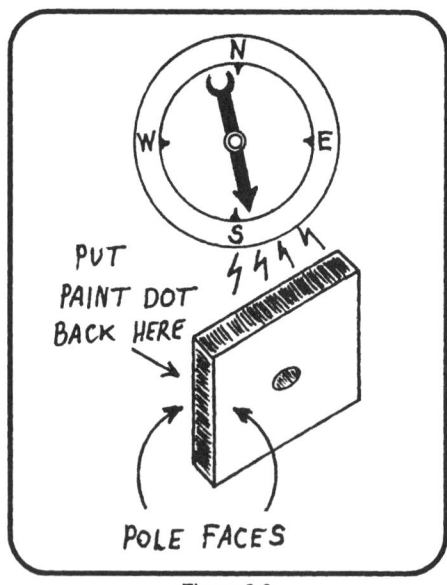

N
W E
S

PUT PAINT DOT BACK HERE

POLE FACES

Figure 6.8
Determining polarity of magnets

manipulating it in the presence of a strong magnet.

The back strap is basically a platform on which to mount the magnets and pole pieces. See figure 6.9. It's fabricated from a segment of steel strap, measuring 2-1/2 inches in length, 1/2 inch in width, and 1/8 inch in thickness. Four holes are located and drilled in the back strap.

The two holes near the center are used to mount the pole pieces. They're sized to pass a 6-32 screw, and are spaced roughly an inch apart. The exact spacing will depend upon the diameter of your coil bobbins. The two holes near the ends of the back strap are sized to pass #6 wood screws. They'll be used later to anchor the back strap to the instrument's wooden base.

The magnets and pole pieces are assembled according to figure 6.10. Each pole piece and its permanent magnet is secured to the back strap with a *brass* 6-32 machine screw. (You can try an aluminum or nylon screw, if you wish. Do not use a steel screw.) The screws poke upward through the back strap, through the magnets, and are threaded into the base of each pole piece.

One magnet is oriented with its blue dot facing upward, the other one's dot faces downward. The screws securing the magnets should be driven until they're snug, but don't overtighten them, or you'll run the risk of crushing the magnets. Once properly assembled, the magnets, the pole pieces, and the back strap were protected with a generous coat of black lacquer.

The last step is to install the electromagnets. The bobbins are slid onto the pole pieces, but you must observe proper polarity. On one pole piece (it doesn't matter which) the coils must be installed so that the blue marker dots all face *upward*. On the other pole piece, the coils must be installed with the blue marker dots facing *downward*.

The purpose of the blue dots and the attention paid to their proper orientation is to ensure that the magnetic fields produced by the coils and the magnets interact in a cooperative fashion.

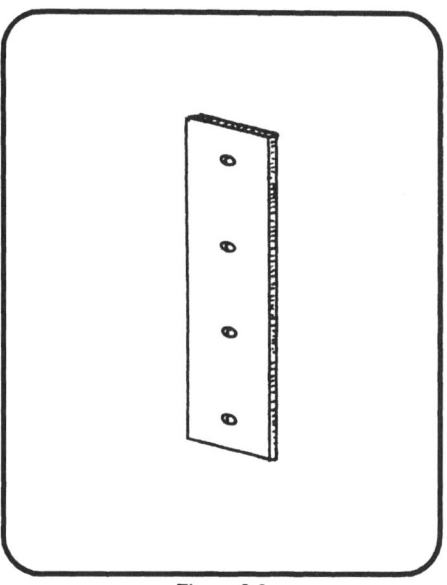

Figure 6.9
Backstrap

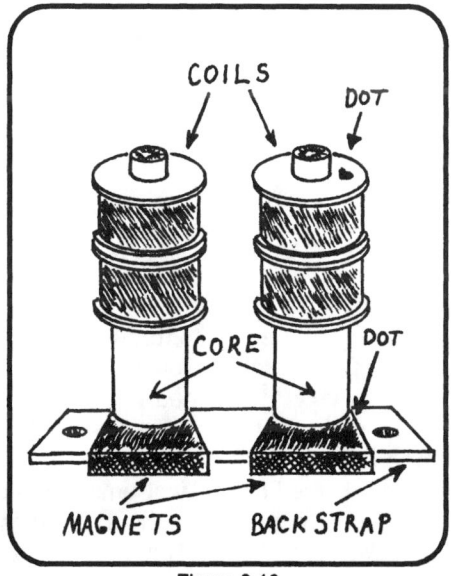

Figure 6.10
Completed driver assembly

Diaphragm / Lower Contact Assembly

In the case of the Brown Amplifier, the magnetic field from the coils and magnets acts upon a delicate metal reed. Mounted on that reed is the lower microphone contact, made of carbon. In developing the Balance-Beam Amplifier, I tried a fair assortment of materials in search of a suitable reed. In the end, I abandoned that approach in favor of a more traditional diaphragm-based design.

The diaphragm was fashioned from the lid of a large-sized can of shoe polish. To prepare the lid, I first cleaned it thoroughly, both inside and out. I then played the flame of a propane torch on the inside face of the lid. The intent was to anneal the metal, as this seems to improve the sensitivity of the instrument. Be warned, however, that the procedure is a delicate one, and requires caution. The metal should be heated just to the point of acquiring a bluish tint, and no further. In a moment of carelessness, the metal can be easily burnt, or warped to the point of uselessness.

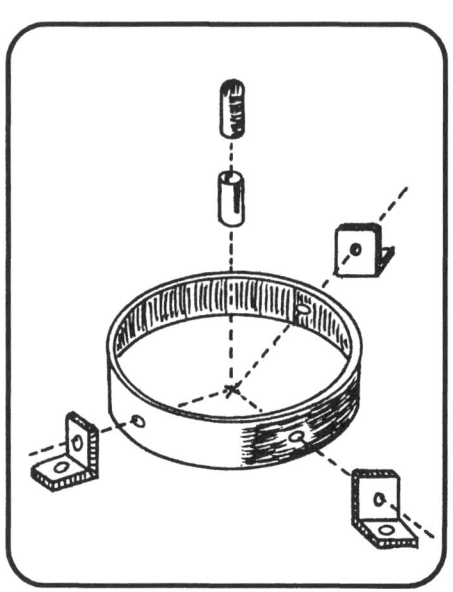

Figure 6.11
Diaphragm and lower contact assembly

Using a strip of paper as a gauge, I subdivided the outer perimeter of the lid into three points. At each point, I drilled a hole to clear a 6-32 machine screw.

Next, I fashioned three "L" brackets of 1/2-inch brass strap, 1/16 inch in thickness. The "L" brackets are drilled and bolted to the periphery of the shoe polish lid.

In the finished instrument, the diaphragm is supported above the electromagnets by a tripod composed of three legs made of 6-32 threaded brass rod. The rods pass through holes drilled in the protruding ends of each "L" bracket, and are secured with a pair of knurled brass thumb nuts. Thumb nuts are available at hardware stores. If you can't find them, 6-32 wing nuts are a good second choice. See figure 6.11.

The lower contact in the Balance-Beam Amplifier was made from a section of carbon rod, which was previously extracted from a dead flashlight battery.

To harvest the carbon rod from a battery, you must first identify a suitable host. In this case, you'll want the plain, inexpensive carbon-zinc type batteries *only.* Do not attempt to dismantle alkaline, lithium, nickel-cadmium, or lead-acid gel cells. The prototype amplifier used the carbon rod from a "D" sized flashlight cell.

Using a hacksaw, cut down the side of the cell, and around each end, as shown in figure 6.12. Exercise extreme caution, as both the saw and the resulting metal edges will be very sharp. When the cutting is complete, the zinc can be split and pulled away, leaving a cylindrical lump of black, pasty chemicals. The carbon rod, of

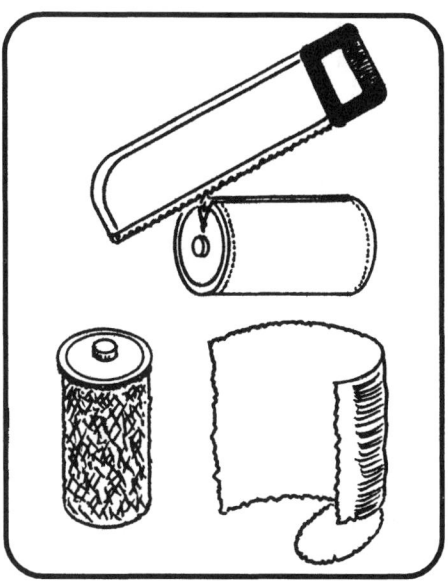

Figure 6.12
Harvesting carbon from old flashlight batteries

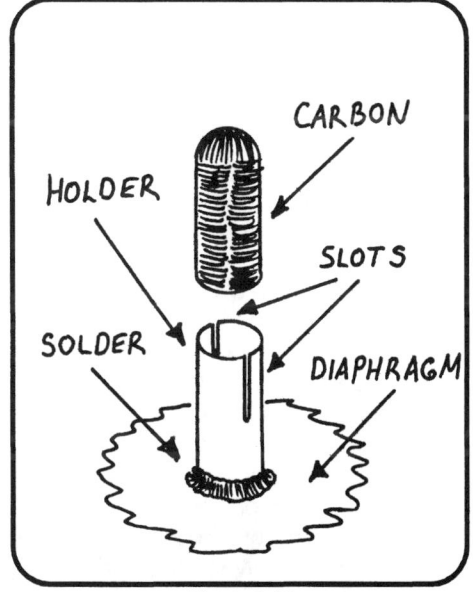

Figure 6.13
Details of the lower contact assembly

course, lies at the center of this mess.

Most of that black paste can be chipped away with a wooden stick or an old screwdriver, but do exercise some restraint. The carbon rod is fragile, and easily broken if stressed. The remainder of the paste can be washed away with water. Consider some eye protection during this work. A pair of latex gloves might not be a bad idea either.

For the record, don't attempt this kind of surgery indoors. The black chemicals in the cell have a wicked habit of scattering everywhere, despite your best intentions to contain them. They can easily ruin clothing, carpeting, or the surface of a kitchen table. You may never find a more certain formula for eviction or divorce, so consider yourself warned.

While the carbon rod may now appear clean, it is actually porous, and has certainly absorbed compounds that could interfere with the optimum function of the amplifier. Donning safety goggles, and holding the rod with a pair of old pliers, I took the added step of baking it in the flame of a propane torch. In fact, I heated the rod to a dull red incandescence, and watched as all sorts of goo percolated to the surface boiled away. Needless to say, this should also be done outdoors, preferably with a light breeze to carry away any potentially noxious fumes.

To shape the electrode, the rod must be loaded gently into the chuck of a hand drill, and spun. As the rod rotates, you can apply sandpaper to give the free end a nice, hemispherical form. When the desired contour has been achieved, continue the operation, using a small swatch of cotton cloth in place of the sandpaper. Applied properly, the cloth will impart an excellent polish to the work.

In the end, you won't need the entire carbon rod, just the polished end and another 1/2 inch or so. To cut the rod, use a small, triangular metal file to make a notch at the appropriate length. With the carbon held in both hands, and the notch facing upward, the rod may now be snapped in two, as you would a pencil.

The carbon contact must now be fastened to the diaphragm. This was done with a simple holder fashioned from a 1/2-inch length of brass tubing. The diameter of the tubing was selected, such that the carbon rod would just slide into it. Using a tiny cutting wheel, two slots were cut into one end of the tube. The tube was then soldered to the diaphragm, with the slotted end pointing up. See figure 6.13.

The slots provide dimensional relief. If the open end of the carbon holder is pinched between the fingers, the opening will be slightly deformed. When the carbon rod is inserted, the holder seizes the carbon and holds it in place with gentle pressure and friction. The appearance of the brass holder with the carbon contact in place has been likened to that of a small pistol cartridge. For those knowledgeable of firearms, an empty shell casing might indeed fit the bill as a holder for your carbon. If you go this route, make absolutely certain that there's no primer in the casing when you go to solder it!

The heat treatment and subsequent soldering operations will probably have flaked much of the paint off of the shoe polish lid. To correct this, I temporarily removed the "L" brackets, and then scoured the whole thing with fine steel wool. Next, I applied a light coat of black spray paint to the outside surfaces of the lid. On the inside, I prefer the bare metal look, so there I applied a thin coat of clear lacquer. To protect the carbon rod during painting, I temporarily removed it, and also wrapped the holder in a layer or two of masking tape. Painting may seem a frivolous step, but don't overlook it. The steel found in "tin" cans is cheap, and notoriously inclined to corrode and rust.

Balance-Beam / Upper Contact Assembly

As I stated earlier, the success of an unstable contact amplifier depends upon the proper adjustment of the carbon contacts, particularly the pressure. In the Brown design, extra coils and circuitry were used to establish and maintain the correct operating pressure between its contacts. Despite a great deal of experimentation, I was never able to replicate the self-adjusting qualities of the Brown device. Instead, I devised a scheme where a delicate balance is used to apply a precise, adjustable pressure to the contacts.

The balance portion of my amplifier is itself composed of three pieces, a head assembly, a pivot mechanism, and a set of counter weights, all of which ride on a balance-beam composed of a 4-1/2-inch length of 6-32 threaded brass rod.

To construct the balance, I began with a piece of 1/4-inch brass rod, which was cut to the length of 1-1/4 inches. Lengthwise, the exact center of the rod was located, drilled through, and tapped for 6-32 threads. The ends of the rod were then turned to sharp points in a lathe. (Note that this operation can also be performed in a drill press with a file.) I call this component the "pivot shaft."

The pivot shaft was threaded onto the balance-beam and advanced until it was located somewhere near the midpoint of the beam. A pair of thumb nuts, tightened from either end of the balance-beam, locked the pivot shaft into position. Geometrically, the balance-

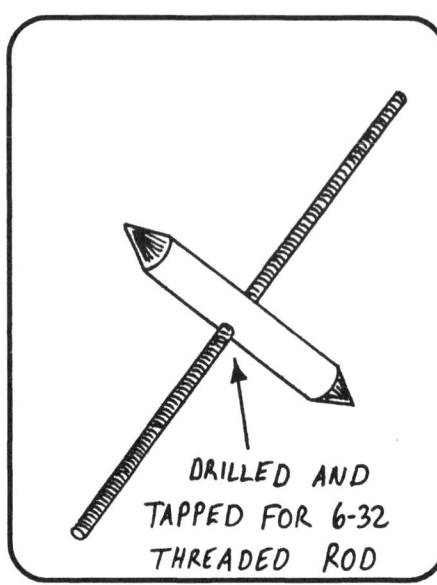

Figure 6.14
Pivot shaft and balance beam

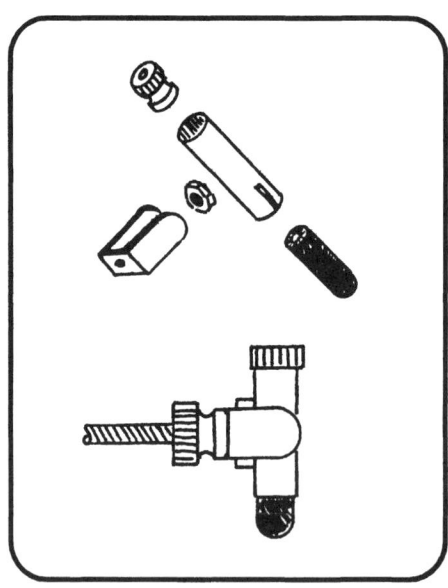

Figure 6.15
Head assembly details

beam and the pivot shaft fit together like the arms of a crucifix, as can be seen in figure 6.14.

The head assembly begins with a brass tube, fashioned and sized to hold a carbon electrode. In this respect, it's quite similar to the carbon holder used on the diaphragm (and described in the last section). See figure 6.15. The top end of the tube was fitted with a knurled brass thumb nut. The nut plugs the top end of the tube and is soldered into place. It provides some mass to the head end of the balance. It can also be fitted with a short screw for additional weight.

A small "U"-shaped bracket, fashioned from brass, and a small hex nut, also of brass, were soldered to the side of the tube. The base of the "U" bracket is drilled to allow access to the threads in the nut. The completed head assembly can then be screwed onto the end of the balance-beam rod. Note that an additional thumb nut, threaded onto the rod, is used as a "jam nut" to lock the head into position. Additional thumb nuts may be added to the head end of the balance-beam, as necessary, to increase the weight at that end.

During my initial experiments, I tinkered with a variety of carbon contacts including pencil leads, graphite, and similar carbonaceous materials. For reasons I can't explain, I determined that contact pairs composed of carbon and graphite seem to work better than contact pairs composed solely of carbon.

In the case of the Balance-Beam Amplifier, the upper electrode is not taken from a battery, but rather, is made from the graphite of an electric motor's commutator "brush." Motor brushes are available at hardware stores and appliance repair shops. The brushes tend to be square in cross section, but can be shaped, polished, and mounted in the same manner as the battery carbons.

The bearing support is a large, "U"-shaped bracket fashioned from 1/2-inch brass strap, 1/16 inch thick. Each leg of the bracket is drilled to clear a 6-32 screw, and is fitted on the inside with a brass hex nut. The hex nuts are soldered into place.

Two bearings were fashioned from 6-32 brass

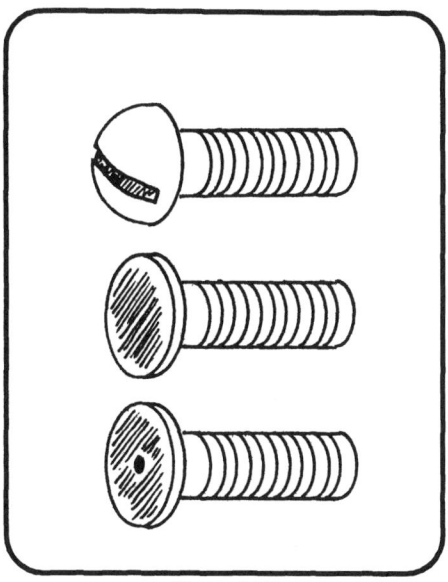

Figure 6.16
Steps in creating pivot bearing screws

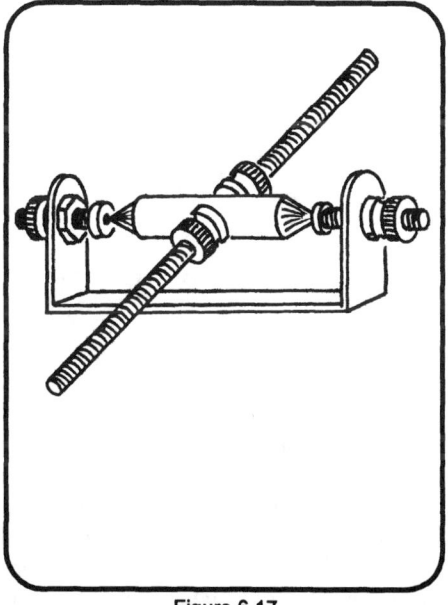

Figure 6.17
Balance beam and completed bearing assembly

Page 57

screws. To make them, the head of each screw is filed down until its slot has vanished, and the head has been reduced to a flat, featureless disk. The center of each disk is subsequently dimpled with a sharp punch, or with a tiny drill bit. See figure 6.16.

The completed bearing screws were then threaded into the bearing support bracket. The flattened and dimpled heads, of course, face inward. To install the pivot shaft, it must be maneuvered so that the points of the shaft will rest in the dimples on the head of each bearing screw. This required some fidgeting, and adjustment of the bearing screws. Once the pivot mechanism is properly adjusted, the pivot shaft and the attached balance-beam virtually "float" between the bearing screws. Any motion of the balance-beam becomes effortless, with no evidence of binding or undue friction.

The bearing screws, having been adjusted, may be locked into place with a couple of brass thumb nuts, threaded on from the outside. See figure 6.17.

Two counter-weights were used on the prototype amplifier. A large weight, composed of a slug of scrap brass, was turned round on my lathe, drilled through, and then tapped with 6-32 threads. It amounts to an oversized nut. It was finished with an old alarm clock gear, which was soldered to its rear face. The smaller of the two weights is simply a brass thumb nut, which was also fitted with a gear. See figure 6.18.

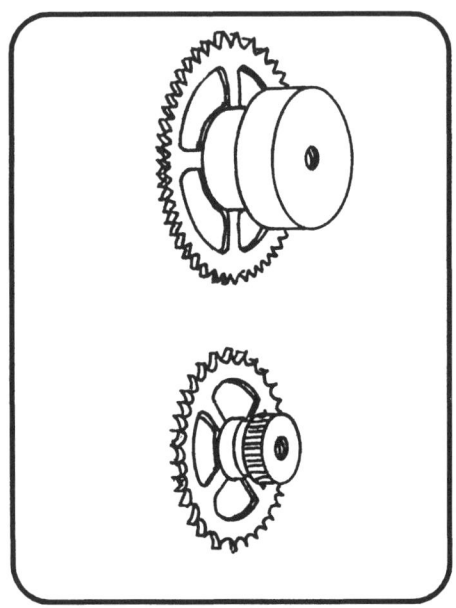

Figure 6.18
Balance counterweights fashioned from brass scrap, thumb nuts, and old clock gears

When completed, the weights were threaded onto the free end of the balance-beam. The equilibrium of the balance mechanism, and thus the pressure on the carbon contacts, can be varied by screwing the counterweights toward the pivot, or away from it. As the weights move toward the fulcrum, pressure on the carbon contacts will increase. As the weights move outward, the pressure will decrease until the beam tips away and the contacts separate.

As the proper balance is achieved, the equilibrium of the system becomes increasingly delicate. The gear teeth on each weight allow fine adjustments to be made without disturbing the instrument too much. The gears can be advanced with the measured poke of a pencil point, or manipulated with the edge of a fingernail. In the finished instrument, the

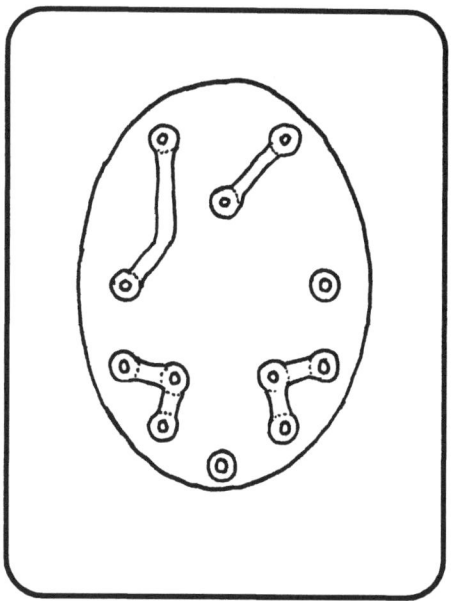

Figure 6.19
The underside of the instrument base

balance-beam assembly must be elevated so that the upper and lower carbon contacts will be in correct alignment. To accomplish this, the bearing support bracket rests on a part I call the "bearing tower." The bearing tower was made from a section of 3/4-inch copper pipe and two pipe caps. Together, they form a cylinder roughly 3 inches in height. The center of each cap was drilled to allow a threaded rod to pass up the center of the tower and protrude slightly from its top.

The rod passes through the base of the bearing support bracket and is secured with a 6-32 thumb nut.

Instrument Base and Terminals

The base of my amplifier was made from a pre-cut, pre-shaped basswood plaque, purchased at a local craft store. It's elliptical in shape, measuring 8-1/2 inches in length, by 6-1/2 inches wide. The wood is about 3/4 inches thick.

The mechanical layout of the instrument should now be evident. The driver assembly was mounted on one half of the elliptical surface with #6 wood screws. The back strap can be screwed directly

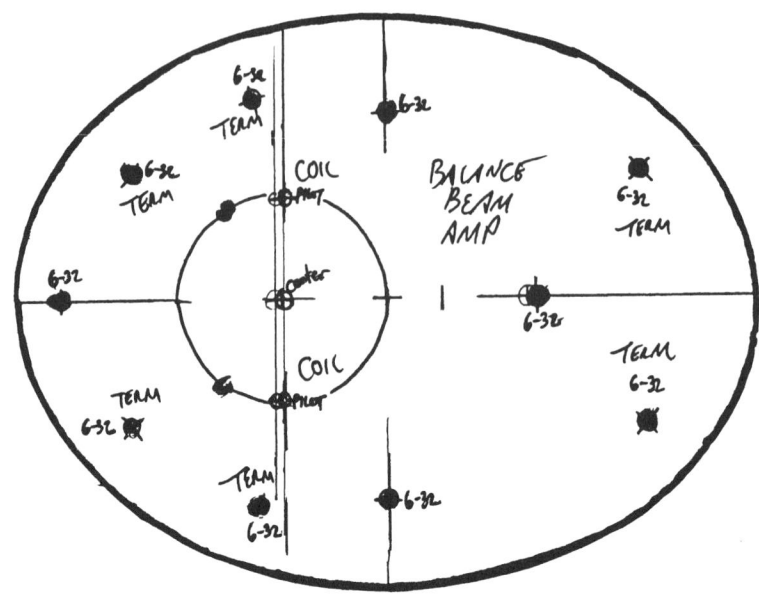

Figure 6.20
An example of a template used to guide drilling

to the plaque, though I used two 1/4 steel hex nuts as spacers to raise the back strap off the surface of the wood.

The diaphragm assembly must be mounted just above the poles of the electromagnets. Three holes must be drilled in the base to accommodate the three threaded brass rods that form the legs of the tripod.

Next, the balance tower must be positioned so that the balance-beam, at rest, will allow the upper graphite contact to rest precisely on top of the lower carbon contact. A hole must be drilled through the base, so that the balance tower can be secured to the base with a length of threaded rod and a couple of nuts.

Figure 6.19 shows the bottom of the base. Note that any screw hole that penetrates the base is countersunk from the bottom with a 5/8-inch spade bit. The countersinking provides ample space for a washer and a nut, while still allowing the base to sit flat on a tabletop. As well, the bottom face of the base features several channels cut into the wood with a router. The channels provide space to string wires from one point of the instrument to another.

Drilling the various holes in their proper location requires planning. One approach is to gather the various mechanical parts, set them on top of the base, align them as desired, and then mark the appropriate positions for holes with a pencil. The drawback to this method is that you tend to leave marks on the wood that must be sanded off later.

I prefer to use a template. A template is simply a piece of paper that represents the wooden base. You can position your parts on it, make measurements, doodle marks, or whatever else it takes to arrive at a final layout. When the template is completed, you tape it to the top surface of the instrument base and drill right through it, using its markings to guide your work. Figure 6.20 shows an image of the template I used to lay out the base for this project.

As a matter of practice, I like to assemble my instruments on raw wood. After I have ascertained that there are no mechanical problems with alignment or fit, I disassemble the device, and proceed to finished the wood.

Finishing the base involves some sanding, the application of suitable stain (I used red mahogany), followed by a couple of coats of satin polyurethane.

Electrical connections are made to the amplifier through a series of six terminals arranged along the periphery of the ellipse. The terminals are composed of brass screws, secured with brass hex nuts, and fitted with knurled brass thumb nuts. To label each terminal, I took the trouble to fashion a set of 3/4-inch brass disks, which were then stamped with an identifying letter "A" through "F." See figure 6.21 for details.

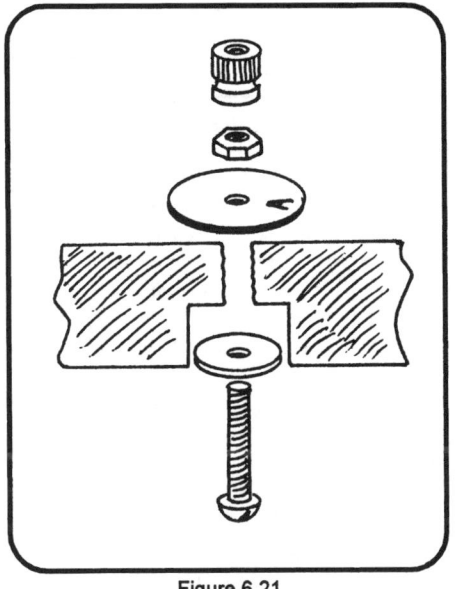

Figure 6.21
Installing terminals

Instrument terminals "E" and "F" provide electrical access to the amplifier's carbon contacts. Terminal "E" is wired to one of the legs supporting the diaphragm. Current can flow up the leg, though the diaphragm, and into the lower carbon contact. Likewise, terminal "F"

Page 61

is wired to the balance tower. Current can flow up the tower, through the balance-beam, and into the upper graphite contact.

Instrument terminals "A" though "D" provide connections to the coils in the driver assembly. Wiring the various coils to these terminals requires attention to the matter of proper polarity. Figure 6.22 depicts the driver coils. Notice they are numbered for convenience. It is assumed the coils #2 and #4 are stacked so that the paint dots (polarity markers) on their respective flanges both point upwards. Coils #1 and #3 are both stacked so that their paint dots point downward. Wiring proceeds according to the following table:

Coil Wiring Details

Coil	Black Wire	White Wire
#1	Connect to terminal "A"	Connect to terminal "B"
#2	Connect to terminal "A"	Connect to terminal "B"
#3	Connect to terminal "D"	Connect to terminal "C"
#4	Connect to terminal "D"	Connect to terminal "C"

Setup and Operation

Once the Balance-Beam Amplifier is complete, the final concern is its proper setup and operation. For the purposes of demonstration, let's revisit the crystal-radio tuner introduced in Chapter 5. We'll replace the microphonic relay with the Balance-Beam Amplifier. See figure 6.23 for the wiring diagram. With a strong local station tuned in, you may hear program material in your headphones immediately.

The first adjustment to make deals with the height of the diaphragm above the poles of the electromagnets. Diaphragm height can be changed by tinkering with the adjustment nuts on the support legs. Ideally, the diaphragm should lie as near as possible to the poles of the electromagnets without actually touching them. You can use feeler gauges to set the gap, though I've had some success using nothing more than a strip of printer paper. Free space can be verified by holding the instrument in front of a desk lamp, and looking for the glint of light between the pole pieces and the diaphragm.

The next adjustment, and the most important, is the equilibrium of the balance itself. The process depends largely on trial and error. Using the edge of a fingernail or the point of a pencil to manipulate the gear teeth, the counterweights are gently rotated and screwed forward or back to affect the balance. Contact pressure is increased as the weights are moved inward, and reduced as the weights move outward.

The objective is to bring the carbons together in firm contact, yet with the lightest possible pressure. Adjustments become successively more difficult as contact pressure is reduced, because the slightest disturbance of the balance will result in chattering noises in the headphones. The process, then, requires a steady hand. When the optimal balance has been achieved, the amplifier will noticeably increase the volume of the program material.

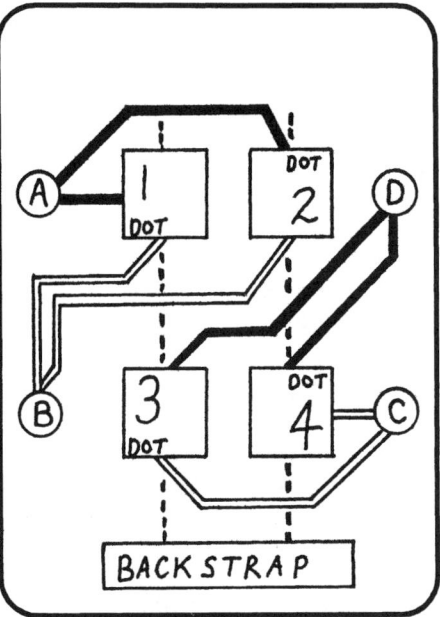

Figure 6.22
How to wire driver coils to the terminal posts

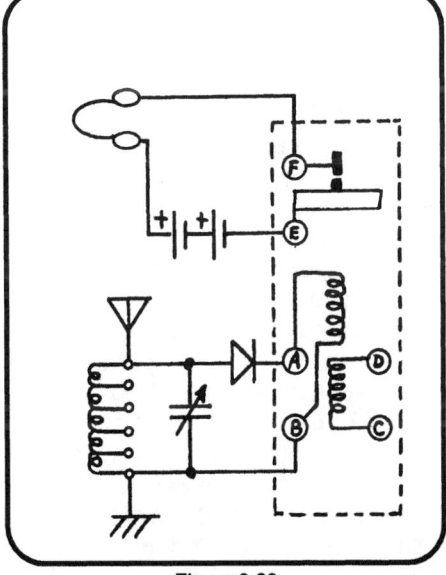

Figure 6.23
A simple, amplified crystal-radio circuit

Page 63

End Results And Going Further

The Balance-Beam Amplifier is an interesting beast with curious idiosyncracies and a bushel-full of surprises.

The design depends upon the vibration of carbon contacts for its operation. It stands to reason, then, that the instrument may be sensitive to external vibrations as well. While the floor of my lab is a thick concrete slab where adequate stability should be a virtual guarantee, I quickly discovered that any agitation of the table surface on which the device rests will result in a flurry of rustling noises in the headphones. This can become quite irritating when you're in the process of tuning or tinkering with a receiver circuit. Ultimately, I separated the amplifier from the tuner, placing them on different table tops.

Tweaked to the point of maximum sensitivity, the carbon contacts really surprised me. After a particularly intricate session of adjustments to the weights, I suddenly became plagued with a rumble in the headphones that waxed and waned at more or less regular intervals. In a flash of insight, I came to realize that my instrument was picking up the seismic rumble of automobiles passing on a roadway *some 300 feet away.* I readjusted the counterweights to increase the contact pressure, and the effect vanished.

Because of this amplifier's sensitivity to external vibration, I can envision an enhanced version fabricated on a marble base. This would significantly increase the mass of the amplifier such that, were it placed upon a thick, compliant, foam rubber pad, it might actually be immune to the various disturbances noted above.

Another unforseen behavior of the amp has to do with the equilibrium of the balance itself when subjected to strong audio signals. During my tests, I tuned in a weaker local station and adjusted the amplifier for maximum volume. The setup worked great until I re-tuned the receiver to a second, more powerful station. The audio was proportionately louder, but it chattered, cutting in and out with momentary gaps of silence. I quickly determined that the stronger station was driving the diaphragm hard enough to actually bounce the carbon contacts apart. Adjusting the weights and increasing the contact pressure solved the problem.

Pressure increases may suppress the skipping problem, but it can also reduce the sensitivity of the amplifier. The real solution to the problem is the addition of some type of damping mechanism to the balance-beam itself. A damper would suppress the beam's tendency to skip, but would still allow for light pressure between the electrodes.

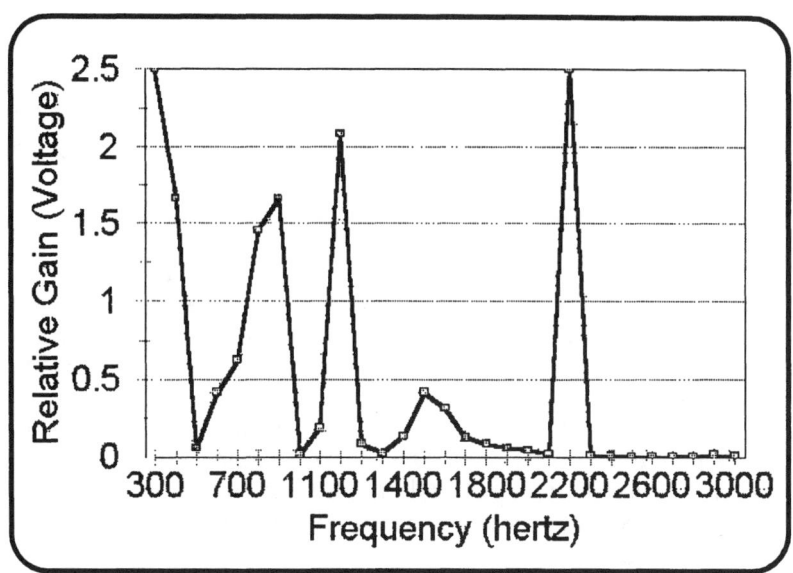

Figure 6.24
Graph showing how the Balance Beam Amplifier's gain changes
as a function of applied frequency

I can think of several ways to implement damping. The most obvious would be the addition of a tiny paddle to the balance-beam itself. The paddle would be immersed in a shallow dish or reservoir filled with mineral oil or some other inert substance with high viscosity. The oil, of course, would resist any rapid movement of the paddle, and thus stabilize the balance.

Needless to say, oil-immersed paddles are troublesome and inconvenient. A clever and a much-improved approach involves the replacement of oil with permanent magnets. In this case, the paddle would be composed of copper, and engineered to swing between the poles of powerful magnets. Like its counterpart in the oil dish, the copper paddle can be moved slowly through the magnets with no consequence. Rapid movement, however, triggers electromagnetic reactions, (eddy currents and so forth) that result in slow, viscous movement of the paddle. Jitter in the balance, thus retarded, would suppress any tendency for the contacts to skip or chatter.

How much gain will the Balance-Beam Amplifier actually provide? That's a fair question. I was curious myself, so I set up a signal generator to drive the coils with a 1-kilohertz sine wave. The input level was set at 1.0 volt. I then measured the output voltage as it fed a set of low-impedance stereo headphones. The voltage there was 1.2 volts. Remember that the gain of an amplifier is defined as its output divided by its input. Thus, in the stated configuration, the Balance-Beam Amplifier provided a total voltage gain of—yes—1.2 times. Given the effort required to build this device, I suspect an anemic gain of 1.2 will produce groans of disappointment from the audience. Before you give up on this design, let's look at the input and output signals a little more closely.

I decided to examine the output of the amplifier to see how much *power* (as opposed to voltage) it was delivering to the headphones. I measured the output current, multiplied that by the voltage, and arrived at a calculated output power value of 27 milliwatts (27 thousandths of a watt).

Next, I performed the same calculation on the input side of the amplifier. I measured the input current flowing from the signal source into the amplifier, and multiplied that by the input voltage. I computed that the amplifier was drawing 0.166 milliwatts of power from the signal source.

The magic appears when you calculate *power* gain instead of voltage gain. Take the output power, and divide it by the input power. You get a gain factor of around 160! In other words, under the stated conditions, the Balance-Beam Amplifier delivered a signal to the headphones that had *160 times* more power than the signal that fed it. Now that's not bad at all!

What's the lesson here? There are different kinds of amplifiers. Some are designed to amplify voltage, some are configured to increase current. Even if an amplifier provides no gain at all, it can still provide a valuable service, namely, isolation. A well designed amplifier will not allow disturbances on its output side to be reflected back to the input.

Recently, I picked up an old 1920's style speaker horn at an auction. On a whim, I attached it to the Balance-Beam Amplifier and drove the amp with a simple crystal-radio set composed of a coil, a capacitor, and a 1N34 germanium diode. Tuned to a strong local station, the horn produced enough music to be heard clearly at the levels of normal conversation. Without the amp, the tuned circuit would have done a poor job of driving that horn directly, so naturally, the power gain is advantageous. What's less obvious is the fact that the horn, connected directly to the tuning components, would have imposed such a burden that the tuner would have lost some of its ability to differentiate between adjacent stations. A tuner in this condition may allow two stations to be heard at once. Instead, the amplifier provided isolation between the delicate tuning circuitry and the burdensome electrical demands of the horn.

While we're on the subject of loading, let's discuss the arrangement of coils on the Balance-Beam amp. Instead of two input terminals, as is the usual case with a speaker, a set of headphones, or that 1920's horn I described, our amp has four terminals. This gives us access to two pairs of driver coils, which can be used and combined in any way you see fit. For example, you may choose to use the first pair exclusively, or alternately, the second set. You can connect the two sets in parallel, and use them that way, or wire them in series. These variations will alter the apparent load that the amp places on any input-signal source. A proper match between a source and a load results in the optimum transfer of energy.

On my prototype, a 1-kHz sine wave, applied to terminals "A" and "B" will see a calculated load of 1,826 ohms. The DC resistance is about 300 ohms. If the signal set is a simple crystal tuner, that's not too bad.

To wire the coils in parallel, terminal "A" must be connected to terminal "D," and terminal "B" must be wired to terminal "C." The signal is applied, as before, to terminals "A" and "B." In this configuration, the 1-kHz signal sees a load of about 1,500 ohms. The direct current resistance here is about 150 ohms. Bear in mind that to parallel the coils, they must be phased properly. Be sure to observe the correct wiring to the specified terminals.

If you connect the coils in series, by connecting "B" to "D," and then applying the signal to "A" and "C," the source will see a load in excess of 6,000 ohms! Series operation is also dependent upon proper phasing. Again, use care in wiring the coils together, or you may find the amp does not work at all. Which configuration is best? That depends on the specific characteristics of your amplifier and what the signal source is.

In the previous chapter, we defined amplification, and through example, established its purpose. The point of amplification is to take an external signal and regenerate it using the energy from a local supply. What's not stated, but always implied, is that the amplifier should perform this task without corrupting or otherwise coloring that signal. The amplifier's output should echo its input in a proportional manner. Nothing should be added to the nature or content of the input, nor should any feature be taken from it.

Consider a hypothetical amplifier with a voltage gain of 5. If we feed that amplifier a 2-volt, 1-kHz sine wave, we expect the output to be 10 volts. If we change the input to 2 volts at 5 kHz, the output should still be 10 volts. In other words, the amplifier should boost all frequencies *equally*, without showing any sort of favoritism or discrimination. If we feed this amplifier a musical symphony, with a rainbow of frequencies and sounds, the fact that the amp is impartial means that the output will be an accurate representation of the input, only larger.

In the real world, no amplifying device is perfect, and no device exists that can process all signals without introducing some manner of distortion. Generally speaking, most amplifying systems are limited in the range of frequencies they can successfully handle.

Human hearing, under good conditions, is capable of detecting sounds ranging in frequency from 20 hertz (cycles per second) to 20,000 hertz. Ideally, an amplifier used to process audio signals should be capable of providing uniform gain for all frequencies between these limits. The span between these limits is called the *bandwidth* of the amplifier. The wider an amplifier's bandwidth, the greater the range of frequencies that it can be relied upon to accurately reproduce.

If you were to drive an amp with a whole series of frequencies between 20 and 20,000 hertz, and in each instance plot the corresponding output, you'd produce a graph depicting the *frequency response* of the amplifier. In the case of a "good" audio amplifier, the test would yield a graph that was essentially straight and flat. In service, this means that all the frequencies within that amplifier's bandwidth receive the same degree of amplification. That's a good thing, because it means that the amplifier will accurately reproduce the signals that are fed to it. By the way, if you've ever heard an audiophile or stereo salesman use the phrase "flat response," this is what they were referring to.

The Balance-Beam Amplifier is *not* a high-fidelity amplifier. To begin with, its useful lower limit seems to be about 300 hertz. If you stimulate it with signals lower than that, the carbon contacts are inclined to chatter. At the top end, it provides little gain above 2 kHz or so. This means that music processed by the Balance-Beam Amplifier tends to sound low and boomy, conspicuous in its absence of any of the sparkle that high frequencies contribute. Note also, that the amp's narrow bandwidth makes the amplification of radio carrier signals (hundreds of thousands of hertz and above) quite impossible.

Worse, between these two limits, the response of the Balance-Beam is anything but flat. Figure 6.24 is plot of the amplifier's response to a series of input frequencies ranging from 300 hertz to 3 kHz. Notice that the response is jagged and irregular. There are certain frequencies that it favors highly, and still others that it chooses to

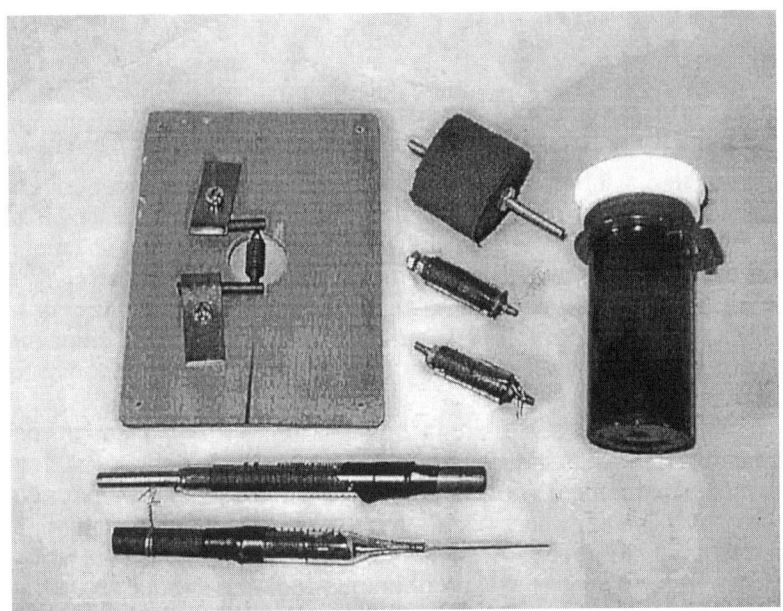

Figure 6.25
An assortment of vibration-sensitive unstable carbon contact devices

ignore or even suppress. This nonuniform behavior is attributable to the fact that the Balance-Beam device is mechanical in nature. In operation, various pieces and parts of the amp are set into motion, so physical factors like mass and elasticity influence the instrument's performance. In the same way that a mechanical pendulum favors a certain period, or a tuning fork favors a specific note, the diaphragm, contacts, balance-beam, and other moving parts each have their own unique resonances.

The Balance-Beam Amplifier is admirable in its existing form, though an ambitious experimenter might seek to extend and or "flatten" the device's response. I suspect there is some room for improvement here without resorting to exotic techniques.

Increasing the mass of the entire balance-beam and adding the damping mechanism might reduce its tendency to chatter, thereby lowering the bottom limit of its frequency response. The high end might be raised by reducing the size and mass of the diaphragm assembly, including the lower carbon contact. Damping materials placed directly on the diaphragm (like tape or thick paint) might help quash unwanted resonant peaks and make the device's response more uniform.

Alternate Carbon Contacts

Before arriving at the final design for the Balance-Beam Amplifier, I conducted numerous experiments with a fair assortment of carbon-based unstable-contact configurations. See figure 6.25. In each case, my intent was to couple them with a set of driver coils to produce a working microphonic relay. Obviously, the most successful scheme was incorporated into the final design, though it doesn't mean that the other approaches lacked merit. In closing this chapter, let me touch on a few of the alternatives I investigated and invite you to experiment with them yourself. Perhaps you'll uncover some detail I've overlooked, to correct the flaws that doomed these experimental devices to the junk drawer.

The first unstable-contact design I tried was fabricated from the carbon rods extracted from some "AA" penlight batteries. First I cut a piece of carbon rod about 3/4 inch in length and used sandpaper to taper each end to a sharp point. Around its waist, I wrapped 8 turns of iron wire.

Next, I cut two pieces of carbon rod, each about an inch in length. Near one end of each piece, I flattened the side with a metal file and then ground a tiny dimple with the point of a small drill bit. Carbon is soft enough that this can be done by applying the bit to the work, and twirling the shaft of the drill between your thumb and forefinger.

The two dimpled pieces were clamped to the face of a small piece of wood using brass scraps. They were arranged to lie parallel, with the dimples facing each other. The points of the iron-clad carbon were inserted into the dimples, and the various parts were adjusted so that the pointed carbon became trapped. It was allowed enough clearance to spin or rattle, but not enough to fall out.

The design had all the trappings of a successful unstable contact. When wired in series with a battery and a set of headphones, any motion of the suspended carbon produced all sorts of noise. The reason I had wrapped it with iron wire is that I had intended to focus the field of the driver coils directly upon it. This meant that I would require neither a reed nor diaphragm. Because of the delicacy of the suspension and the low mass of the suspended carbon, I had expected this to be an extremely sensitive amplifier. Instead, I was never able to generate anything more than crackling sounds.

Another experimental device was based loosely on the design of some vintage telephone equipment. It consisted of two disks made of graphite, oriented face to face, with a 1/4-inch gap between them.

The disks were wrapped in a belt of fabric, which was glued to the perimeter of each disk. The space between the disks was filled with fragments of crushed carbon rod. Note in the photo that a mounting screw appears from each end. One end of the capsule was intended to be secured to a fixed anchor, and the other to a diaphragm or similar structure for the electromagnets to act upon.

Bolted to the base of a soup can as part of an initial test, I wired it with the requisite battery and headphones. Yelling into the open end of the can, I found that the capsule actually functioned admirably as a crude microphone. It proved unsuitable in an amplifier design because of its excessive mass, and the tendency of the carbon particles to congeal and pack together. Graphite can be harvested in nice chunks from the crucibles used by jewelry makers to melt small quantities of gold or silver.

A third microphone was fabricated from the glass body of an eye dropper. A short length of brass wire, wound into a tiny pancake spiral at one end, was inserted into the dropper. The free end of the wire protruded from the point of the dropper, and the spiral prevented the wire from falling out. A small quantity of crushed carbon was loaded into the dropper, and the open end was plugged with a section of carbon rod.

Mechanically, the carbon was intended to be fixed or anchored. The brass wire was supposed to be attached to a reed or diaphragm to be acted upon by the driver magnets. The arrangement was extremely sensitive to movement at certain rare moments, and useless the bulk of the time. I abandoned it because of its unpredictability.

A similar incarnation of the glass-bodied carbon capsule used two carbon rods to seal the ends of the tube. One, of course, was fixed and glued into place. The other was carefully sanded and polished to fit and slide effortlessly within the bore of the glass. The space between them was filled with crushed carbon. Note that the carbon plugs at each end were fitted with 4-40 mounting screws. The glued plug was intended to be mounted to a fixed surface, and the loose plug bolted to a reed or diaphragm.

The function of this capsule was initially disappointing until I dumped out my homemade crushed carbon and used the granules extracted from an actual telephone mike. The grains removed from the commercial mikes appear to be polished. In the glass capsule, they functioned exceptionally well.

I did make one attempt to polish my homemade carbon grains. To do this, I mixed equal parts of crushed carbon to common table salt, and then loaded the mixture into an empty plastic medicine vial. I shook the mixture for an extended period of time, and then examined the carbon grains beneath a microscope. They had been significantly smoothed by the abrasive action of the salt grains. Separation would have involved mixing the powder with water, dumping it into a paper coffee filter, and rinsing the carbon trapped in the filter. Left in the sun to dry, the grains could then be loaded in the glass capsule.

I eventually abandoned this design because carbon particles had the bad habit of working their way around the side of the moving carbon plug, causing both leakage and the tendency for it to seize.

The final design I entertained was yet another glass capsule. It featured a fixed brass plug at one end, and a moveable brass plunger in the other. Instead of filling the intermediate space with crushed carbon, I experimented with a stack of tiny carbon disks. Using a jeweler's saw, I sliced the carbon rod from a "D" flashlight battery into a series of disks, each about 3/32 inch in thickness. The disks were washed in lacquer thinner, and then stacked inside of the glass tube. I think I used about 25 disks between the brass electrodes.

As microphone or vibration sensor, the contraption was a dismal failure. The stack represents a lot of mass, a lot of friction, and poor mechanical coupling from one disk to the next. Vibrations just don't pass down the stack very well. However, I did notice that if the stack was connected to a battery and small lightbulb, I could easily vary the intensity of the bulb with a surprising degree of precision. This stack of carbon disks now resides in the attic of my mind, waiting for the moment that I need a sound design for a homebrew variable resistor.

References

Buchsbaum, Walter H., *Buchsbaum's Complete Handbook of Practical Electronic Reference Data,* copyright 1978, Prentice-Hall, Inc., Englewood Cliffs, N.J., pp. 22-23, (wire gauge data)

Bucher, Elmer E., *Practical Wireless Telegraphy,* copyright 1921, Wireless Press Inc., New York, pp. 169-171, (details of the Brown Amplification Relay)

Friedrichs, H. Peter, *The Voice of the Crystal - How To Build Working Receiver Components Entirely From Scratch,* copyright 1999, H. P. Friedrichs, ISBN 0-9671905-0-9, pp. 36-39, (details on salvaging wire, tips and winding electromagnets), pp. 119-122, (miscellaneous information about wire and coil winding)

Ghirardi, Alfred A., *Radio Physics Course,* copyright 1933, Radio & Technical Publishing Company, New York City, New York, p. 224 (information on inductive reactance), pp. 686-692, (speaker driver information)

Grob, Bernard, *Grob Basic Electronics - Seventh Edition,* copyright 1992, McGraw-Hill, ISBN 0-02-800762-X, New York, New York, pp. 730-731, (voltage, current and power gain, definitions and mathematics)

Chapter 7
The Needle Box Transformer

An interesting behavior of the Balance-Beam Amplifier is its apparent preference for one type of headset over another. Initially, when I set up the device and began playing with it, I used a set of comfortable, low-impedance stereo headphones. There was no particular reason for the choice, they just happened to be handy.

Eventually, my thoughts turned to the high-impedance set I normally wear when I'm playing with crystal-radio sets. Usually, when wired directly to a crystal tuner, the high-impedance set outperforms anything else around. In the case of the Balance-Beam Amplifier, the use of the "good" headphones actually degraded the amplifier's performance. There is probably more than one explanation for this behavior, but it furnishes an excellent reason to touch upon the ins and outs of properly *matching* signal sources to their loads.

What is matching? Usually, when a signal source is connected to a load, we'd like to ensure the maximum transfer of power. Energy sources have unique characteristics, as do various loads. If a source and a load are somehow incompatible, the degree of power transfer will be poor. If the source and load have the correct attributes, the transfer of power can be significantly improved.

Consider, for a moment, the engine in an automobile. The crankshaft on the engine turns quite fast. The engine can produce a fair amount of torque (twisting force), but a lot of rpm (rotations per minute). What do you want at the tires? Nearly the opposite, i.e., a fair amount of rpm, but a lot of torque, particularly when you're accelerating from a stoplight. The characteristics of the engine do not match the requirements of the wheels. This is why you never see automobiles in which the engine is linked directly to the tires. In the

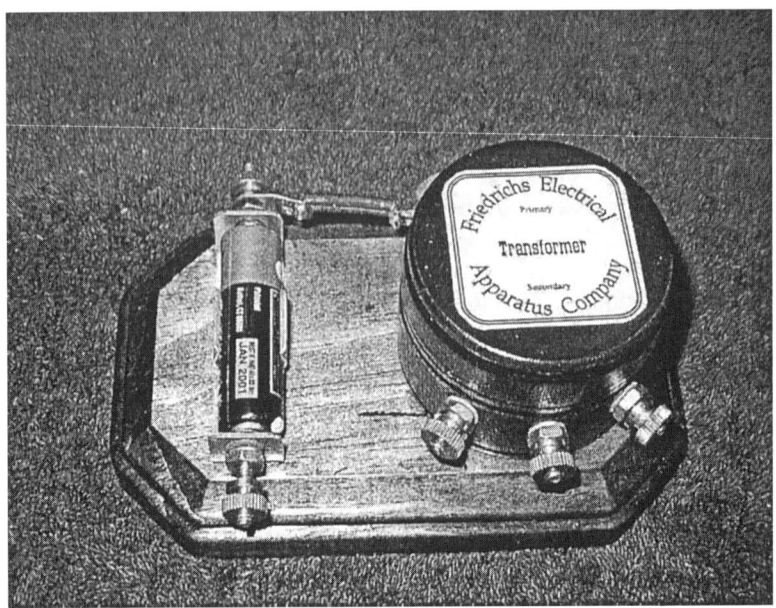

Figure 7.1
The Needle Box Transformer

parlance of our amplifier studies, we'd say that the power source (the engine) and the tires (the load) are not properly matched.

Enter the transmission. The transmission is a mechanical conversion device. It helps convert rpm to torque. Inserted between the engine and the tires, both are suddenly satisfied. The engine spins quickly, thinking that it's driving a high-speed, low-torque load. The tires are happy, because they think they're being driven by a slow-turning, high-torque engine. The transmission acts as in intermediary to link the two together, each on their best terms. It has the capacity to match the engine to the wheels.

All of us have seen historical photos of early bicycles. The early designs featured crank pedals attached directly to the shaft on the front wheel. Over time, this design proved to be unsatisfactory, because the characteristics of the power source (your legs) did not match well with the drive requirements of the load (the wheel).

If you now examine a contemporary bicycle, you'll note that the crank pedals are attached to a large sprocket linked via a chain to a smaller sprocket affixed to the rear wheel of the bike. Slow, high-torque motion of the pedals produces fast, lower torque motion of the rear wheel. Once again a conversion device, this time the sprockets and chain, matches the source to the load in order to improve the transfer of power. With the large sprocket on the crank, your legs can move at a comfortably slow pace while the rear wheel is driven at a much higher speed.

In the world of electronics, sources of energy and loads also have specific characteristics. If the two are properly matched, optimal power transfer may occur. If they aren't matched, you can run into the sorts of trouble that I observed with my high-impedance headphones.

A transformer is an electronic component that can be used, among other applications, to perform this type of matching. Like the transmission in a car, it's a conversion device that's able to simultaneously satisfy the requirements of both ends of the circuit. After pondering my observations, I eventually designed a simple transformer for use with the Balance-Beam Amplifier and found that it greatly improved the instrument's performance when used with a high impedance-headset. In this chapter, I'd like to describe the construction of a primitive matching device I call The Needle Box Transformer. See figure 7.1. While it's not, strictly speaking, an instrument of amplification, its application stretches from unstable contact amplifiers, to vacuum tubes, transistors, and beyond.

A More Detailed Explanation

As I stated in the introduction, I'm disinclined to launch into complicated mathematics for it's own sake. The auto transmission analogy above is not a bad one, and if it provides a suitable explanation for you, go ahead and skip this section. If you're the curious type who prefers more detail, try to follow along as I lead you through a specific line of reasoning.

Let's begin with a flashlight battery and a simple thought-experiment. Suppose we take that battery and purposefully short-circuit it, i.e., apply a thick piece of copper wire to connect the positive terminal to the negative. *(Warning: Don't try this for real. At best, you will have ruined a perfectly good battery. At worst, you may be burned or injured if the casing of the cell should rupture.)*

If we start with Ohm's law and juggle the terms around to solve for current, we get the expression in Figure 7.2. Basically, the current (I) in that wire should equal the battery voltage (V) divided by the resistance (R) in the wire. We know that the cell produces 1.5 volts, and we're assuming that the resistance of the wire is very, very low. If you run this equation on a calculator, you'll find that the predicted

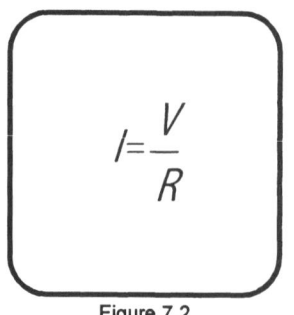

$$I = \frac{V}{R}$$

Figure 7.2
A variation of Ohm's law

battery current skyrockets. The closer the wire's resistance value is to zero, the higher the current climbs. In reality, there's no such thing as a zero-ohm wire, and mathematically, you're never permitted to divide by zero, but it quickly becomes evident that with a perfect conductor, the current should zoom to infinity.

Needless to say, the real-life current will never get that high. In fact, I actually conducted this test on a brand-name alkaline "D" cell using appropriate equipment and found the current topped out at 6 amps. That's a far cry from infinity! What happened?

Let's go back to Ohm's law with this new information. First, we juggle the terms around to solve for resistance, and then we plug in the known information. See figure 7.3. The voltage is 1.5 volts, and the current during the short circuit condition was 6 amps. R turns out to be 0.25 ohm. Hmm...the circuit behaves as though there is a quarter-ohm resistor in the loop somewhere...but where?

It turns out that it's *hidden* in the battery itself! The physical attributes of the cell-its geometry, materials, and chemistry-result in reactions that produce an electric current with the baked-in equivalent of a one-quarter-ohm resistor. On the basis of Thevenin's Theorem (a system to simplify electrical networks for the purpose of analysis,) we can model the real battery as shown in figure 7.4. The model consists of an ideal voltage source of 1.5 volts in series with a resistor measuring 0.25 ohm. If you short-circuit the model with a fictional piece of zero-ohm wire, you'll find that the current will max out at 6 amps just like the real battery. Try the math.

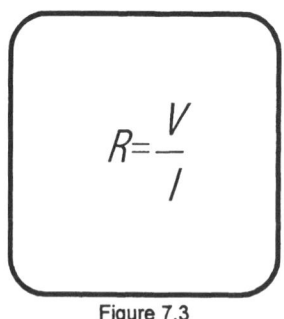

$$R = \frac{V}{I}$$

Figure 7.3
Another variation of Ohm's law

A load is the electrical burden placed upon a signal source. That load might be the input wires to some other electronics, it might be a speaker, or it might be lamp or a motor. The most basic load is a simple resistor. If we return to our model of the flashlight cell, and connect it to a load resistor, electrical power will transfer from the source (the battery) to the load (the resistor). The load, in turn, will respond by doing what any resistor does with electric current. It converts that power into heat. The question is this: What should the resistance value of the load be to ensure the maximum transfer of power?

Let's begin with a load resistor value that's smaller than the battery's source resistance. For argument's sake, we'll use a value of 0.10 ohm. Using Ohm's law, we can calculate the current in the load by dividing the voltage of the source (1.5 volts) by the sum of the source and load resistors (0.35 ohm). This results in a calculated current flow of 4.3 amps. See figure 7.5. The power (P) dissipated by a load resistor is defined in figure 7.6 as the square of the current times the load resistance value. If you square 4.3 amps and then multiply it by 0.10 ohms, you get about 1.8 watts.

Next, we'll try a load resistor that's greater than the source impedance. This time, for the sake of example, we'll use a value of 0.5 ohm. Once more, we divide the source voltage by the sum of the resistance resulting in a computed current of 2 amps. Square that current, and multiply it by the load resistance (0.5 ohm) and you get a power value of 2 watts.

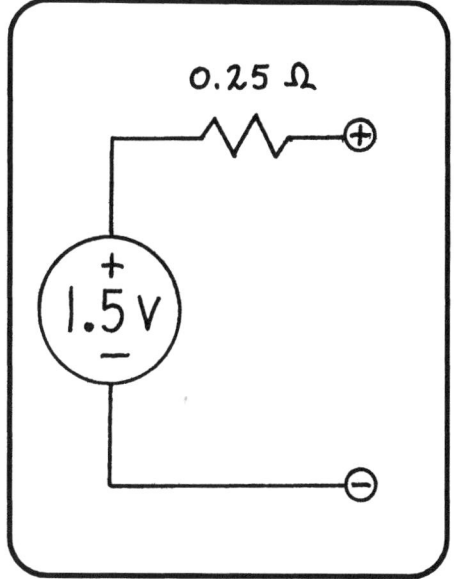

Figure 7.4
A behavioral model for the battery

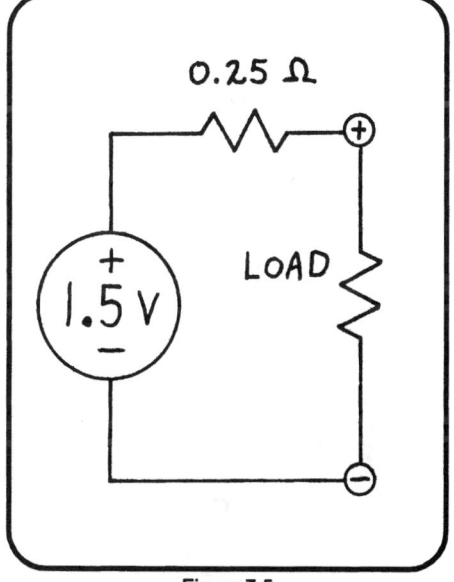

Figure 7.5
The battery with a load

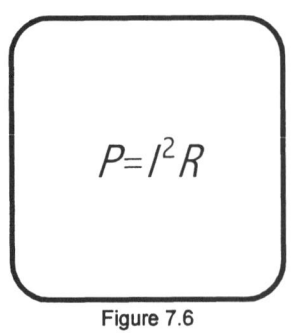

$$P = I^2 R$$

Figure 7.6
Calculating power

Finally, let's make the load resistance *equal* the source impedance. In this case, both will be 0.25 ohm. If you repeat the calculations above, you end up with a final power value of 2.25 watts. You can repeat the calculations as often as you'd like, using whatever load resistance values you want. In the end, you'll discover that the maximum transfer of power will occur only when the load resistance matches the battery's internal resistance.

The battery and the resistors described above represent a specific example of a signal source and load. In a broader sense, it's worth understanding that all electronic signal sources have, hidden within them, some degree of internal resistance. If you use Thevenin's Theorem to analyze these sources, like we did with the battery, they can be modeled in a similar fashion.

Of course, not all signal sources and loads are direct-current devices. It's a fair bet that most of them, including amplifying circuits, process and or utilize alternating currents. In the world of AC, the effects of a device's internal capacitance or inductance may be significant. Capacitors and inductors have the ability to discriminate between signals of varying frequencies, altering their effective resistance (reactance) in response to the composition of the signal. This complicates the Thevenin model. Because these frequency-sensitive components must be accounted for, the model may contain additional inductors and/or capacitors as well.

While the term *resistance* tends to be

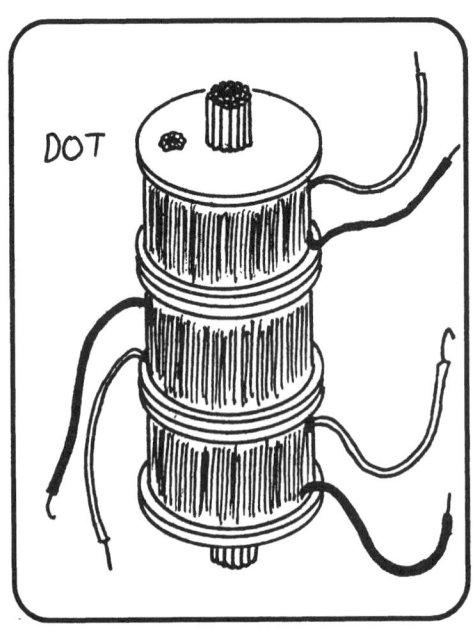

Figure 7.7
A stack of coils on an iron wire core

Page 80

used in conjunction with direct-current circuits or devices that are not frequency sensitive, the word *impedance* embraces the current-impeding qualities of both resistors and frequency-sensitive devices like coils and capacitors. Thus, the phrase *source impedance* is more generic than *source resistance,* because it represents the model's built-in current impeding components, regardless of whether they are resistive, inductive, or capacitive in nature.

This brings me to the purpose of matching. We discovered that maximum power transfer between a source and a load occurs when the source and load impedances are identical. If the source impedance is known, achieving maximum transfer is a simple matter of selecting an appropriate load.

Unfortunately, there are many situations where the source and load impedances are already defined and unchangeable. In addition, they may be different enough that the transfer of power through a direct connection would be exceptionally poor. Power transfer can be improved by the insertion of a matching device, like a transformer, between the source and the load.

Building the Transformer

A transformer is a magnetic conversion device, useful for numerous applications including source-to-load matching. Structurally, it consists of two coils, a so-called *primary* and a *secondary*. Both are wound on a common magnetic core.

A signal is fed into the primary coil, which creates a magnetic field in the core. This field induces a current in the secondary coil, thereby transferring energy from one to the other. The nature of the transfer from primary to secondary depends, in part, on the number of turns of wire in each coil.

This is not unlike the behavior of two gears, meshed together. If the "primary" gear is driven by a motor, the rotational behavior of the "secondary" gear will depend upon the ratio of the diameters of the two.

The coils for the Needle Box Transformer were fabricated in the same way as the coils in the Balance-Beam Amplifier. The coil forms are clear, plastic sewing-machine bobbins wound with fine wire. The coils must be wound, terminated with flexible insulated leads, and then tested and marked for polarity. Refer to Chapter 6 for specific details. Two such coils will be required. These coils will later be stacked to form the transformer's secondary.

The primary coil consists of another plastic bobbin, this time wound with a slightly heavier gauge of enameled magnet wire. I used #26 gauge wire, salvaged from an old solenoid valve. Like the other coils, the winding process can be greatly streamlined if an electric drill is used to spin the bobbin. Unlike the other coils, no special termination is required, because #26 wire is thick enough to be handled without the danger of breakage. Just leave a few inches of each lead dangling.

The core was composed of a bundle of iron wire. I located a spool of wire at a hardware store and snipped off a few dozen segments, each about 1-3/4 inches long. The wires were carefully heated to a dull red in a propane flame, and then set aside to cool slowly. The intent here was to soften and anneal the wire. Each wire received a light coat of lacquer from a spray-paint can.

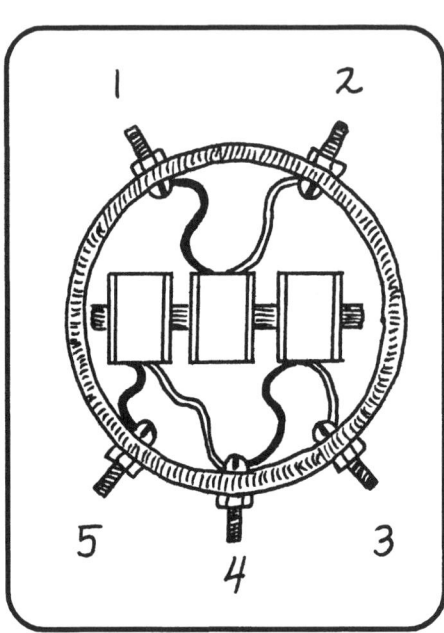

Figure 7.8
Transformer coils and their terminals

To assemble the transformer, the three bobbins must first be stacked in the proper sequence. Begin with one of the secondary coils. Place it on the bench top in front of you, with the blue polarity dot facing upward. Next, lay the primary coil on top. The primary coil's polarity is unimportant. Finally, add the third bobbin to the stack, making sure that its blue dot also faces upward. This results in a coil stack wherein the primary coil is sandwiched between the two secondary coils.

Once the bobbins were properly sequenced, I began inserting wire segments into the hollow center of the bobbins until it became packed. I finally stopped when no more wire could be inserted into the core. See figure 7.7.

Figure 7.9
The inside of the Needle Box Transformer

The peculiar name of the transformer stems from its housing, a small, round, wooden container intended to store pins and needles. It was purchased at a local craft store for a couple of dollars. Five holes were drilled around the perimeter of the transformer to accept 6-32 brass screws to act as binding posts. Each screw was fitted with a solder lug and inserted though the wall of the box from the inside. On the outside, the screw was fitted with a brass washer and nut. Each binding post was finished with a knurled brass thumb nut, available at your local hardware store.

I also drilled a small hole in the center of the base of the needle box to accept a 6-32 tee nut. I figured the tee nut would come in handy when it later came time to mount the transformer onto another surface.

Once the mechanical fitting was complete, I removed all of the hardware from the needle box, sanded it, stained it, and then coated it with polyurethane. When it was thoroughly dry, I reinstalled the posts and other hardware.

The next step was to wire up the binding posts. To do this, the coil assembly was laid upon the table with the core in a left-to-right

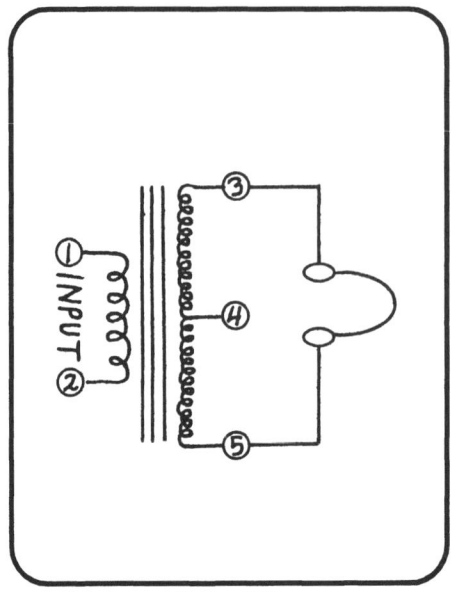

Figure 7.10
Using the transformer with headphones

orientation. Starting with the leftmost coil in the stack, I located its leads, soldering the black one to terminal "5" and the white to terminal "4." The leads to the middle coil were soldered to terminals "1" and "2" with no regard to polarity. Finally, I located the leads to the rightmost coil, soldering the black lead to terminal "4" and the white to terminal "3." See figure 7.8.

Once the internal wiring was complete, it was time to install the core into the housing. I cut a small disk of plastic to lay on the floor of the needle box. This was intended to provide some shielding to the coils in the event that a mounting screw was driven too far into the base of the needle box. Then, I carefully placed the coil assembly inside the box, taking care to see that none of the wires were pinched. With everything settled inside the housing, I took the added step of stabilizing the structure with beeswax. The wax was melted in a double boiler composed of two tin cans and poured carefully into the needle box. (Never heat wax directly on a stove; you're liable to start a fire.)

$$\left(\frac{N_S}{N_P}\right)^2 = \frac{R_S}{R_P}$$

Figure 7.11
Computing the turns ratio to achieve a given impedance match

Figure 7.9 is an image of the interior of the transformer prior to the inclusion of wax.

I took the additional step to mount the transformer to a stained, wooden base, and fitted that base with a battery holder. Technically, the battery has nothing to do with the transformer, though together, the whole unit makes a nice accessory to go with the Balance-Beam Amplifier.

As per my usual practice, I took the aesthetic liberty to fabricate an old-style label for the top of the transformer, identifying its terminals.

End Results and Going Further

Figure 7.10 depicts a schematic for hooking up the transformer. In this case, we're matching the low-impedance characteristics of the Balance-Beam's output to the high- impedance characteristics of the "good" headphones. Happily, the transformer works as advertised.

I made some measurements and then calculated the input impedance of the primary to be about 12 ohms (stimulated by a 1-kHz sine wave). The output, measured between terminals "3" and "5," ranged around 8,500 ohms. The beauty of terminal "4" is that it's a center tap. If you'd prefer less output impedance, use terminals "3" and "4" instead. In the latter case, I computed the impedance to be about 3,500 ohms.

In the Needle Box Transformer, we know that the secondary has many turns of wire, the primary, comparatively few. The primary favors low impedances, the secondary favors high impedances. As it turns out, there is a definite relationship between the ratio of turns (from primary to secondary), and the resulting ratio in impedances (from primary to secondary). Figure 7.11 shows the equation. This expression is most useful when we know the source and load impedance that we wish to match. It becomes a trivial matter to then compute the necessary turns ratio. A suitable transformer may then be designed.

This Needle Box Transformer is primitive in its use of a straight, open magnetic core. The open core has the undesirable property of allowing its magnetic field to radiate outward from its ends. This has the potential to cause interference to other devices. Likewise, the open core makes the

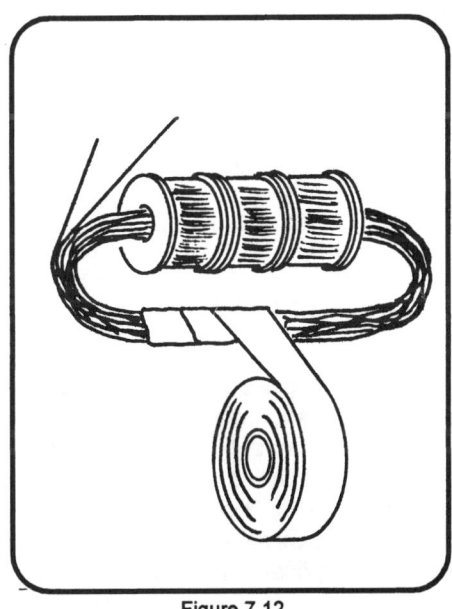

Figure 7.12
A closed core composed of iron wire

transformer highly susceptible to the induction of electrical noises from external sources.

To address these concerns, nearly all modern transformers are fabricated with closed cores. Their cores are fashioned to pass through the coils, bend around, and join themselves in a closed loop. The closed loop helps to contain the internal magnetic field, and offers increased immunity to the effects of external magnetic disturbances.

It is possible to fabricate a closed core with iron wire, it just takes a little more effort. To do so, the iron wire segments must be cut to a somewhat greater length. The wires are threaded through the core, and the free ends are bent into the shape of a closed letter "C". Additional wires are added to the core until it's full. The gaps where the free ends of each wire meet should be staggered. The core can then be stabilized by wrapping it with tape. See figure 7.12.

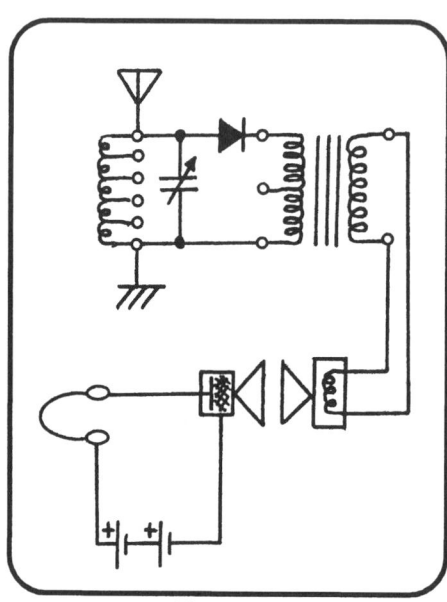

Figure 7.13
Using the Needle Box Transformer with amplifying relays

The Needle Box Transformer can be used in other configurations with equal success. While it was conceived to translate the low-impedance output of the Balance-Beam Amplifier to a suitable drive for high-impedance headphones, its capabilities are both symmetrical and reversible. In other words, it can be connected backwards to help translate a high-impedance source to a lower impedance load.

In Chapter 5, we lamented the fact that the microphonic relays built from telephone parts have a comparatively low input impedance, which makes them difficult to interface to crystal tuners. By employing the Needle Box as shown in figure 7.13, the match between the tuner and the amplifier is improved, resulting in sharper tuning and greater volume.

The Needle Box Transformer has one final attribute that, while unrelated to amplification, may inspire some additional experimentation. The ratio of the number of windings between the primary and secondary coils makes the Needle Box instrument an excellent step-up transformer. Properly applied to the primary coil, 1.5 volts can trigger the production of 90 volts or more at the terminals of the secondary! This is enough, for example, to light a neon bulb, and high enough to be felt as a distinct tingle in the fingertips if they are applied to the output terminals.

Transformers are really intended to be used with, and process, alternating currents. A pure direct current, like that from a battery, connected directly to the transformer's primary will yield nothing at the secondary. In order for the transformer to function, the input signal must fluctuate.

In the case of the Balance-Beam Amplifier, the battery current was made to fluctuate through the action of the vibrating carbon contacts. If you want to use the Needle Box to generate high voltage, you can vary the primary current with an "interrupter" composed of a metal file and a nail. Figure 7.14 shows the Needle Box

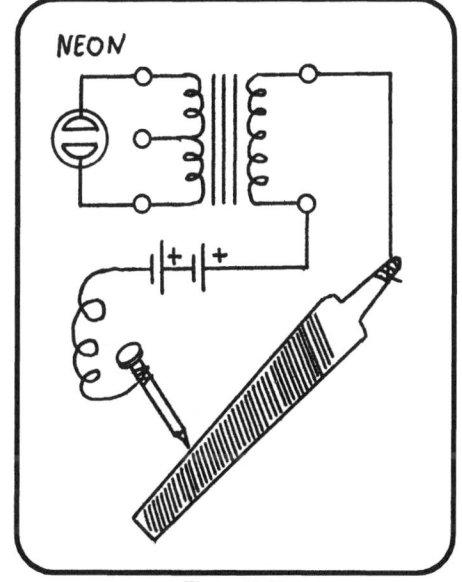

Figure 7.14
Creating high voltage with the
Needle Box Transformer

Transformer wired to light up a neon bulb. To do it, simply drag the nail across the face of the file. Every time the point of the nail skips from one tooth of the file to the next, the primary current is momentarily interrupted, causing a high-voltage spike to appear at the transformer's secondary winding.

Resist the temptation to use more than 3 volts or so as a power supply when using the transformer in this fashion. Higher primary voltages will indeed raise the secondary voltages, but at a risk.

Page 87

This transformer was not designed to handle excessively high voltages. If an internal spark occurs, the coil insulation may be damaged, resulting in a short-circuit. At that point, the transformer will be ruined. The burned coils will have to be repaired or replaced.

References

Buchsbaum, Walter H., *Buchsbaum's Complete Handbook of Practical Electronic Reference Data,* copyright 1978, Prentice-Hall, Inc., Englewood Cliffs, N.J., pp. 22-23, (wire gauge data), pp. 130-131, (transformer construction,) p. 137 (impedance matching transformers)

Friedrichs, H. Peter, *The Voice of the Crystal - How to Build Working Receiver Components Entirely From Scratch,* copyright 1999, H. P. Friedrichs, ISBN 0-9671905-0-9, pp. 91-92 (melting and application of beeswax as a potting material)

Ghirardi, Alfred A., *Radio Physics Course - An Elementary Text Book on Electricity and Radio,* copyright 1933, Radio & Technical Publishing Co., New York, New York, pp. 655-657, (mathematics and transformer design for impedance matching)

Grob, Bernard, *Grob Basic Electronics - Seventh Edition,* copyright 1992, McGraw-Hill, ISBN 0-02-800762-X, New York, New York, pp. 220-222, (Thevenin's Theorem), pp. 298-299, (source to load power transfer)

Henny, Keith, *Principles of Radio,* copyright 1945, John Wiley & Sons, Inc., New York, New York, pp. 102-103, (using step up transformers with direct current)

Henny, Keith, *The Radio Engineering Handbook,* copyright 1941, McGraw-Hill Book Company, Inc., New York, New York, pp. 380-380, (considerations on the design of impedance matching transformers)

Morgan, Alfred Powell, *Wireless Telegraph Construction,* copyright 1913, D. Van Nostrand Company, New York, New York, p. 38, (preparation of induction-coil cores)

Chapter 8
The Vacuum Tube

The instruments discussed thus far are, at best, modest solutions to the amplification problem. While they offer measurable improvement to the performance of otherwise passive devices like the crystal tuner, they tend to be unstable, exhibiting an undesirable sensitivity to external vibration. Perhaps worst of all, the mechanical nature of instruments like the "sound intensifier" or "microphonic relay" means that their ability to respond to rapidly changing signals is inherently limited.

There are any number of signals whose frequencies range far in excess of those reachable by mechanical contrivances. The carriers used in ordinary broadcast AM radio, for example, oscillate at frequencies *five hundred times* greater than the highest frequency the Balance-Beam Amplifier can process. While the latter is passable as an audio amplifier, it's completely useless as a tool to amplify the much higher frequency radio waves. For this reason, radio communication technology could only reach its full potential with the introduction of a practical, multipurpose amplifying device. The vacuum tube answered this need, triggering a revolution in radio and electronics as a whole.

While the vacuum tube is generally considered to be the product of 20th-century ingenuity, the phenomena central to its operation have been observed and recognized perhaps for 200 years or more. Early experiments with the *electroscope*, for example, suggested that heated bodies of metal have unique electrical properties.

An electroscope is a primitive instrument for detecting electric charges. Typically, it consists of a glass bottle fitted with a rubber stopper. A wire passes through the stopper. Outside, the wire is fitted with a terminal composed of a metal ball or disk. Inside the bottle, the wire is bent into an "L"-shape. Suspended from the "L" are two exceedingly thin, delicate leaves of gold. Under normal conditions, the leaves each hang vertically.

Electric charges can exert influence over one another in a manner not unlike the interaction of magnetic poles. As is the case with magnetism, like charges repel, while opposite charges attract each other. If an electrically charged object, like a comb, is brought in contact with the terminal of the electroscope, the charge travels into the bottle and electrifies the gold leaves. Since both leaves are fed by the same wire, they both accumulate the same type of charge. Remember, like charges repel, so the leaves are forced apart, and remain spread like an inverted "V." Under ideal conditions, the leaves will remain apart indefinitely, or at least until the charge is somehow neutralized. See figure 8.1.

Figure 8.1
The Leyden Jar
(From *Electricity and Magnetism*, 1905)

In 1873, an experimenter by the name of Guthrie discovered that the leaves of the electroscope could be made to collapse if a metal ball, heated to a dull red, was brought near the electroscope's terminal. Somehow, without even coming into contact with the electroscope itself, the heated ball was able to influence the charge on the gold leaves.

Ten years later, in 1883, Edison struggled with the development of his electrical light. At the time, his lamps were fabricated with carbon filaments. Many of the filaments developed hot spots associated with a diminished cross section at those points. The hot spots not only degraded the life of the lamp by contributing to the premature failure of the filament, they also spewed particles of carbon, which condensed on the inner surface of the bulb and blackened it.

As part of an investigation to correct the problem, Edison introduced a metal plate into the glass bulb. The plate was connected through a galvanometer (a type of meter that detects feeble currents) to the positive terminal of the filament supply. To his surprise, he detected the flow of an electric current. When the galvanometer was instead terminated at the negative terminal, nothing happened. Somehow, in the first instance, electricity flowed though the evacuated space between the heated filament and the cold plate. Busy as he was, Edison never invested significant effort studying this strange phenomenon. The so-called *Edison Effect* would remain a mystery for more than a decade.

J. A. Fleming, of the University of London, had knowledge of Edison's discovery, and of similar work conducted in 1885 by a fellow named Preece. Fleming recognized that the rectifying properties of Edison's modified lamp made it ideal for use as the detector of radio signals. In 1905, Fleming patented his own version of the vacuum-tube diode, intended for radio telegraphy, which

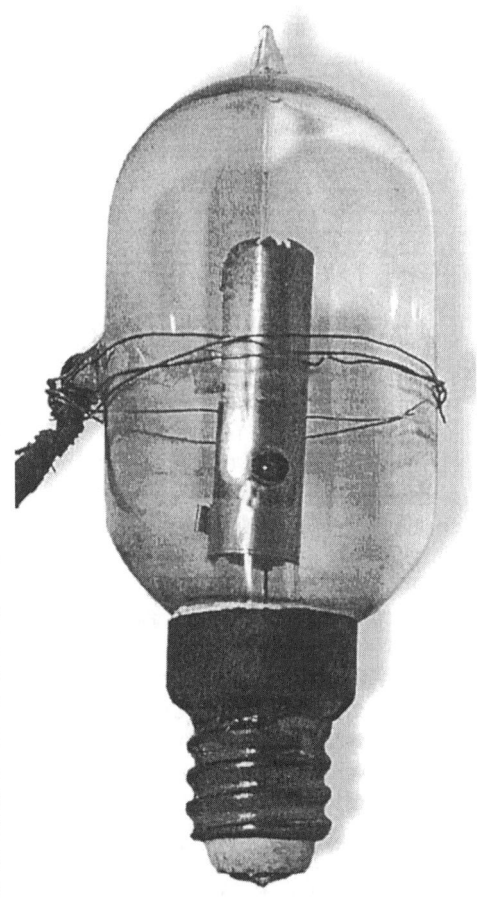

Figure 8.2
The Fleming Valve
(Courtesy of John Jenkins
www.sparkmuseum.com)

eventually became known as the *Fleming Valve*. See figure 8.2.

Finally, in 1907, Lee DeForest introduced a metal mesh or grid between the filament and the plate. By varying the electric charge on the grid, it became possible to regulate and meter the flow of current from the filament to the plate. This modification proved monumental. With it, a feeble external signal could control a local power supply. (This phrase should now sound familiar!) The three-element tube, containing a filament, grid and plate, was called a *triode*. The triode was the first real, fully electronic amplifying device.

The Nature of the Atom

The introduction above lends some insight into how vacuum tubes came into existence, and possibly some clues as to their internal construction. In order to understand precisely *how* they work, one must turn to the atom.

Ancient Greek philosophers pondered the nature of matter, postulating that it might be composed of invisible, microscopic particles. The word *atom* is derived from their term *atomos,* meaning "uncut." This is in reference to the supposed indivisibility of these elementary units of matter. As late as 1900, many scientists clung to the notion that atoms were solid, indestructible particles. Due to the ceaseless efforts of numerous physics pioneers, it eventually became evident that the atom, in fact, was composed of smaller, elementary bodies. Over time, three subatomic particles were identified and characterized, the *proton*, the *neutron*, and the *electron*.

On an atomic scale, the proton is a comparatively large particle that carries a *positive* charge. Both the mass and charge of the proton have been measured and quantified, but for many purposes these numbers are too clumsy to use. For convenience, scientists often refer to the proton as a standard unto itself, with an arbitrary mass of (1) and a charge of (+1).

The neutron is similar is some respects to the proton. It has roughly the same mass, so like the proton, it's considered to have a mass of (1). Electrically speaking, neutrons are *neutral*, i.e., they have no electric charge whatsoever. Thus, they are said to carry a charge of (0). The electrical neutrality of the neutron means that, as a rule, it's never a factor in any ordinary electrical or chemical process.

Electrons are different animals entirely. To begin with, they're extremely light, about 1/1800th the mass of a proton. For practical

Particle	Mass	Charge
Proton	1	+1
Neutron	1	0
Electron	~0	-1

Figure 8.3
Elementary particles and their characteristics

purposes, their mass is considered to be (0). They do, however, carry a significant electrical charge. The charge is identical in magnitude to the proton's, but it's *negative* instead of positive. The electron's charge is said to be (-1). See figure 8.3.

As knowledge of these constituent particles began to accumulate, physicists wondered how they fit together to make up the atom. Among the most popular models was that proposed by Niels Bohr in 1913. Over the decades since, Bohr's model has been revised, and to some extent superceded, as the atom has turned out to be a much more complicated and extraordinary structure than anyone could have imagined. Protons and neutrons, for example, have since been found to consist of a bizarre menagerie of even tinier fragments. Electrons exhibit an almost schizophrenic ability to act as discrete particles under one set of conditions, while appearing to all the world as waves under the next. These details not withstanding, the Bohr model paints a mental image that's easily grasped, making it adequate for our introductory purposes.

In essence, the Bohr model is planetary in nature. Central to the model is a sun-like body called the nucleus. The nucleus is merely a group of some number of protons and neutrons clumped together. Electrons are said to whirl about the nucleus in varying orbits like so many tiny planets.

To "build" a hypothetical atom, we start with the nucleus. The simplest, most primitive nucleus consists of nothing more than a single proton. A proton, as you recall, carries a (+1) electric charge. If, we place a single electron in orbit around this nucleus, the electron's (-1) charge precisely cancels the proton's positive charge for a net atomic charge of zero. This yields a nice, neutral atom. What about mass? The proton has a mass of one unit, and the electron is considered to have a mass of zero, for a total atomic mass of one. The atom we've just described is called *hydrogen*.

Helium is slightly more complex. Its nucleus consists of two protons and two neutrons. In terms of electric charges, each proton contributes a charge of (+1), while the neutrons contribute nothing. The total charge in the nucleus, then, is (+2). In this case, two electrons must orbit the nucleus. The sum of their negative charges cancels the positive charges in the nucleus yielding another balanced atom.

Helium is a much heavier atom than hydrogen. As we said at the outset, the nucleus contains two protons and two neutrons. Each of these particles is said to have a mass of one unit (and the electrons are said to have no appreciable mass), making the total atomic mass of helium, four. See figure 8.4.

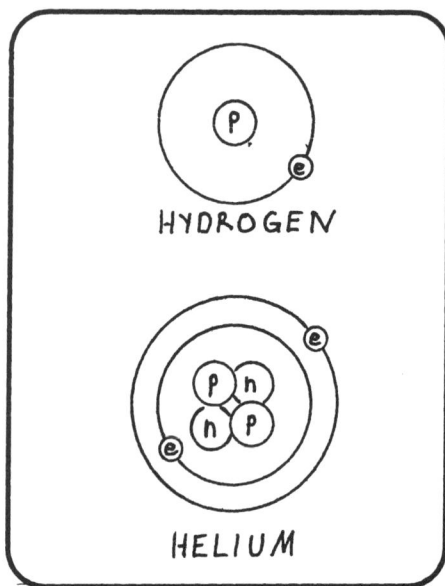

Figure 8.4
The construction of hydrogen and helium atoms

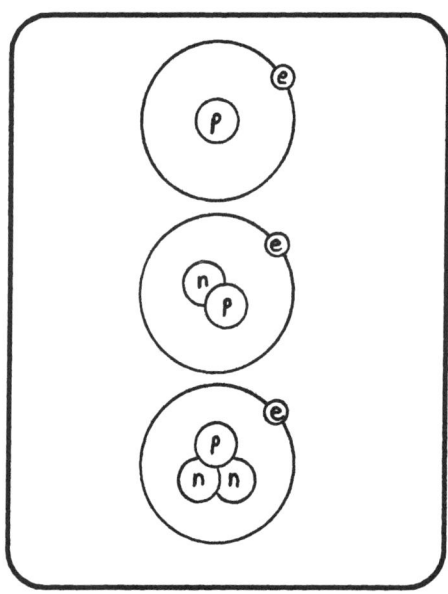

Figure 8.5
Isotopes of hydrogen

Since neutrons have no charge, it's reasonable to wonder what purpose they serve in the nucleus. The answer is complicated, and its details are still the subject of some debate. Bear in mind, however, that all protons carry the same positive charge. Clumped together in the nucleus, they might well be expected to repel each other and fly apart. There is some evidence that the presence of neutrons in the nucleus serves as a "glue" to hold the nuclear structure together. In fact, the more protons present in the nucleus, the more neutrons one tends to find.

An atom with three protons in the nucleus is called *lithium*. The lithium nucleus is stabilized with four neutrons for a total atomic mass of seven. Three negatively charged electrons orbit this nucleus, one to cancel the charge of each of the three positive protons.

If we extend the discussion to include beryllium, boron, carbon, nitrogen, and so on, we can draw a number of conclusions. First, every element is composed of atoms with a unique number of protons. Put another way, *the number of protons in the nucleus of an atom determines its chemical identity.* An atom with one proton in its nucleus will always be a hydrogen atom. Two protons will always yield helium. Eight protons will always give you oxygen. All copper atoms contain twenty-nine protons, and so forth.

Scientists sometimes reference atoms by their *atomic number.* The atomic number of an element is simply the number of protons in its nucleus. The atomic number of hydrogen is 1, the atomic number of helium is 2, and so on. The atomic number zinc is 30. On the basis of this, one immediately knows that a zinc atom must contain thirty protons in its nucleus. There are over 100 elements known to exist, each with its own unique atomic number. The largest atomic number has slowly risen over time as new, artificially created, elements have been added to the list.

While we're on the subject of the nucleus, let's diverge for just a moment to consider a theoretical experiment: Suppose, for the sake of argument, that we create three atoms. Each atom is built with a single proton in its nucleus, and a single electron in orbit. If you understood the essence of the paragraphs above, you'll immediately recognize these atoms as hydrogen.

Now, let's make this interesting. The first atom we'll leave untouched. We'll add a neutron to the nucleus of the next atom. To the last atom, we'll add two neutrons. The modified atoms are depicted in figure 8.5. What is the result of this tinkering?

To begin with, we can easily see that the atomic mass of each of the three atoms must now differ. However, we stated earlier that the identify of an atom depends only on the number of protons in its nucleus. Since each atom in this experiment still contains but a single proton, each atom must still be hydrogen! On considering this, you might reasonably conclude that hydrogen comes in three "flavors," with atomic masses of 1, 2 or 3. It does! Scientists call these varieties *isotopes.* Isotopes are atoms that share the same atomic number (number of protons), but differ in their atomic mass.

If we now turn our attention outward, away from the nucleus, we eventually arrive in the domain of the electrons. Here, the structure of the atom is equally intriguing.

The atoms described above are presumed to be electrically balanced. For the total charge in the atom to equal zero, every positively charged proton in the nucleus must be mathematically canceled by a negatively charged electron somewhere in orbit. It follows, then, that the atomic number for an atom not only indicates the number of protons in the nucleus, it tells you the total number of electrons in orbit as well.

The notion that atoms are constructed as miniature solar systems, complete with electrons in orbit like tiny planets, is appealing and even romantic. It's also misleading. In our solar system, the rules governing the trajectories of orbital bodies are fairly lax. There are no theoretical barriers against "planets" moving through any given region of space as long as certain simple laws of gravity and motion are satisfied. Bohr recognized that, in the case of electrons, the rules were much more stringent.

In contrast, he proposed that the orbital motion of electrons was actually limited to a well-defined set of standard orbits. Under this system, an electron may travel in one allowable orbit or another, but never above, below, or in between. By way of analogy, one might think of these rules in the terms of a nested set of circular highways. On the road, painted lines define precisely where travel is permitted. Note that driving on the shoulder, the median, or straddling lanes is strictly forbidden.

Orbits depicted on paper are necessarily two-dimensional. They're usually represented with closed curves like circles or ellipses. Bear in mind that real atoms are *three-dimensional* creatures. They have height, width, and depth. Likewise, the orbital motion of their electrons is also three dimensional. They whirl about the nucleus with such rapidity that they actually become blurred or smeared, defining their orbital envelope as a surface, like the shell of an egg. Scientists, in fact, speak of the various permissible orbits as *shells*. Shells are numbered, 1, 2, 3, and so forth, starting with the shell closest to the nucleus.

Additional rules define how many electrons may fill each shell. The first shell, for example, cannot contain more than two electrons, no matter what. The second shell can't support more than eight. The third shell can't hold more than eighteen electrons, and so on. As if that wasn't complicated enough, all of the shells beyond the first contain sub-shells. These sub-shells, identified with the letters *s*, *p*, *d*, and *f*, are like multiple lanes within a larger highway. Still more rules, outside of the scope of this discussion, determine how the electrons are distributed among these lanes, and in what order they may be filled.

In the automotive world, lane changes on a busy highway typically require additional energy. A car in the slow lane of a highway will probably have to speed up in order to move successfully into the passing lane. When the driver decides to return, he'll likely shed that additional energy, slowing to his previous velocity in order to merge back into the slow lane.

The electrons in Bohr's atomic model are similarly capable of "lane changes." Under proper conditions, they may snap from one allowable orbit to a higher orbit, but only if they are sufficiently stimulated by an external source of energy. Electrons that have been excited to jump to a higher level will not remain there indefinitely. The electron in its excited state is like a pyramid resting on its point. The condition is abnormal and unstable. The tendency is for the electron to tumble back into the orbit in which it belongs. What happens to the energy that raised the electron to the higher state? It's released as a photon, and may be seen as a flash of light.

By the way, if you've ever studied a neon sign in operation and marveled at its eerie glow, you're already familiar with this process. The high-voltage electric current that flows through the tube agitates the neon atoms inside. Some of the electrons are stimulated to snap to higher orbits where they linger for an instant before tumbling back to their preferred orbits. As each electron falls, it releases a packet of light energy, cumulatively producing a ghostly luminescence.

Most illustrations depicting the atom's internals, i.e., the nucleus and orbiting electrons, tend to portray the atom as a fairly compact structure. In truth, these pictures are intended to show only the structural relationship between the various parts. They're never drawn to scale. This begs the question, "how far do the electrons orbit from the nucleus?"

The Handbook of Physics credits Ernest Rutherford as having determined the ratio of the nuclear radius to the atomic radius to be on the order of 1×10^{-5}. What does this mean? Well, if our goal is to draw a hydrogen atom to scale, and we make the nucleus the size of a dime, the electron must be drawn in orbit nearly a half mile away! Needless to say, this presents some practical problems for the illustrator of a physics book.

Scaled another way, if we imagine that the hydrogen nucleus to be the size of a small beach ball, its orbiting electron will be found at the lonely distance of 14 miles! So, what lies in the vast region between them? Absolutely *nothing*! By far, most of the volume of any atom consists of empty, barren, vacuous space. This is true whether the matter in question is a wisp of gas, or the heaviest bar of gold. The next time you drive a nail with a hammer and fail to control your swing, amuse yourself with the fact that your thumb will have been bruised and bloodied by an encounter with virtual nothingness!

The relatively large distances between the nucleus and an atom's electrons becomes relevant when external forces come into play. Depending on the atom and precisely what the external stimulus is, some of the outer-orbit electrons may actually be torn free and swept away!

An electric current is constituted by the flow of electrons through a given material. Materials whose outer electrons are easily yanked free and manipulated are called *conductors*. Silver, copper, and aluminum are excellent examples of this. Materials that cling jealously to their electrons are called *insulators*. Insulators are far less willing to relinquish their electrons, unless they're placed under extreme duress.

If you'd care to experiment with some of this, the scientific apparatus required includes a carpet, a pair of fuzzy pink rabbit slippers, and a dry day. Put on the slippers, shuffle your feet across the room, and touch a doorknob. You'll probably hear a pronounced snap as an electric spark jumps between your finger and the knob. In this case, the friction between your slippers and the carpet has torn electrons from the carpet's fibers, which then collect and pool in your body.

When you touch the doorknob, you provide a path through which the electric potential can be neutralized. Your body releases the captured electrons to which it's not entitled, and the electrons are afforded the opportunity to migrate back into the matter from which they were taken. Electrical balance is thus restored.

Another way to tear electrons from atoms is through the application of heat. All materials contain some quantity of heat, which sets atoms and molecules into vibratory motion. If extreme heat is applied, certain compounds, metals in particular, will begin to shed electrons. If the temperature is raised high enough, the electrons may even leap from the surface of the material into the surrounding air. This process is called *thermionic emission.*

Let's revisit the electroscope, mentioned near the beginning of this chapter, and reconsider Mr. Guthrie's observations. First, the electroscope is given a positive charge. This causes the gold leaves to be repelled and separated. Next, a metal ball, heated to incandescence, is brought into the proximity of the electroscope. We now know that intense heat can cause some materials to shed electrons through the process of thermionic emission. Negatively charged electrons emitted from the heated ball pass through the air, and strike the electroscope's terminal. Slowly, the emitted electrons begin to cancel and neutralize the positive charge previously held in the electroscope. When enough electrons have been transferred, and the net charge in the scope has been reduced to zero, its leaves will collapse. The mysterious behavior of the electroscope suddenly makes sense.

The Thermionic Diode

An electron leaping from the glowing surface of a metal globe faces obstacles that may not be evident at first glance. The gases that comprise our atmosphere, for instance, don't seem to form a particularly robust or substantive medium. However, if you lower the window of a fast-moving car and extend the palm of your hand into the slipstream, it quickly becomes evident that ordinary air can exert a profound mechanical influence on much more tangible bodies. Anyone who's ever suffered the company of a tornado or hurricane will testify to this.

The mixture of gases called "air" is all around us. Once an electron has been ejected from a heated surface, it'll probably encounter a molecule of one of those gases. The resulting collision will halt the electron's exodus, and it may well be deflected back to the material from whence it came. The average distance the particle can travel before striking another, or *mean free path*, is extremely small.

To visualize this type of interference, picture yourself trying to run a straight path through the plaza of a crowded shopping mall. You may advance some short distance, but the odds are that, somewhere along your course, you'll eventually collide with another patron. If the other shopper is large enough, he may well send *you* back from whence *you* came!

The heart of Edison's lightbulb was a delicate filament made of carbon. An electric current passing through the filament heated it to incandescence. The problem with a heated lamp filament is that, in open air, it's quickly destroyed. Oxygen from the atmosphere combines with white-hot carbon, and the filament is actually burned away. Edison sealed his filaments in blown glass bulbs, and then connected those bulbs to a vacuum pump in order to remove as much air as physically possible. He reasoned that a filament can't burn if there's no oxygen left in the bulb.

While the reduction in air pressure within a lamp serves to lengthen the life of the filament, it also produces an interesting and important side effect. Reducing the number of gas molecules floating inside the bulb assists the escape of electrons from the filament,

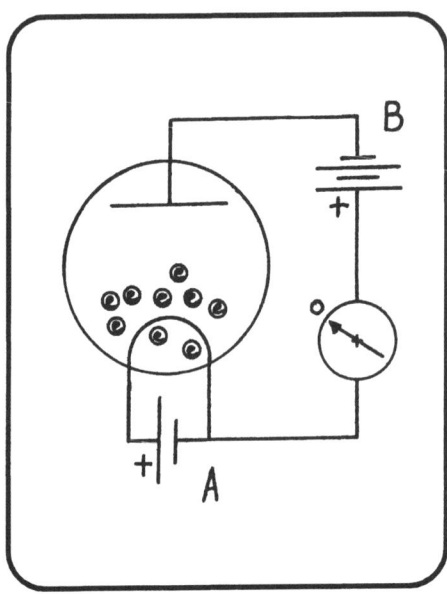

Figure 8.6
Thermionic diode in the *off* state

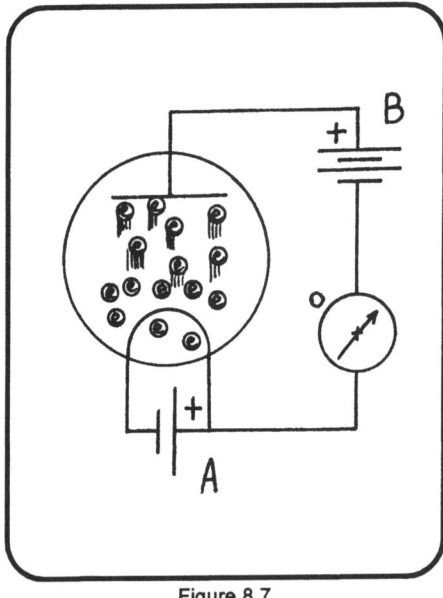

Figure 8.7
Thermionic diode in the *on* state

because the electrons are less inclined to strike and be repelled by various gas molecules. The reduction in gas molecules serves to increase the mean free path of the emitted electrons.

In ordinary air, an electron can travel no farther than a few millionths of an inch before striking a gas molecule. If the pressure within a glass bulb is reduced by a factor of a thousand, the same electron can travel roughly a thousand times farther. In the end, the higher the degree of vacuum achieved, the fewer gas molecules remain to obstruct the motion of ejected electrons. In terms of our shopping mall analogy, an evacuated bulb is like the mall after closing time. A quick run across the plaza becomes trivial, as there are no longer any customers to bump into.

To some extent, the number of electrons emitted from the filament is dependent upon temperature. The hotter the filament, the more agitated the atoms become, and the more inclined they are to release electrons. This emission isn't boundless, however. Freed electrons tend to pool in a charged cloud around the filament. Engineers call this the *space charge*. Once the space charge is established, any additional electrons seeking their freedom will find their escape thwarted by their brethren floating in the space charge. Remember, like charges repel each other, so the electrons already floating in the cloud are inclined to repel newcomers.

The addition of a metal plate to the interior of the bulb gives it some interesting properties, and some important practical applications. Figure 8.6 shows a bulb of the type just described, featuring a filament and a plate sealed in a vacuum. A power source, which we'll refer to as the "A" battery, is connected to the filament in order to light it up. A second battery, referred to as the "B" battery, is used to apply a charge to the plate. A current-measuring device called an *ammeter* is connected in series with the "B" battery to measure the quantity of electricity flowing from the battery.

Electrons flow from the negative terminal of the battery, through the ammeter, to the plate sealed within the bulb. The trouble is, they can't go any farther. To begin with, the plate isn't hot enough to emit electrons,

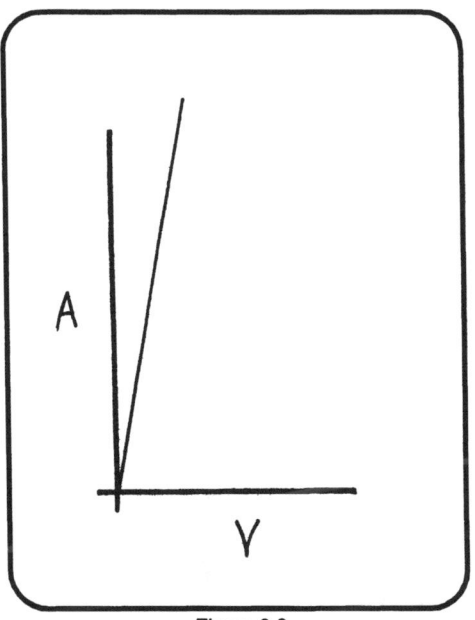

Figure 8.8
Current flowing through a wire, as a function of applied voltage

and even if it were, there's no reason for them to migrate through the vacuum to the filament. The filament is hot, and it's surrounded by a negative space charge. If, by some chance, an electron were emitted from the plate, it would surely be turned back. The net result is that no appreciable current flows in the "B" battery circuit.

Now let's reverse the "B" battery, as shown in figure 8.7. This time electrons, supplied by the battery, are fed to the filament. The filament is heated, so it's capable of accepting those electrons and emitting them into the cloud around the filament. The plate has a positive charge, so it attracts electrons from the cloud, which migrate through the vacuum, into the plate. From the plate, the electrons travel back to the battery. Note that in this case, a complete electric circuit is established, and the real flow of electric current becomes visible on the ammeter.

In summary, if the plate is positive, current can flow through the bulb from the filament to the plate. If the plate is negative, current cannot flow. This is the essence of the effect that Edison witnessed. This one-way-only behavior is the combined product of thermionic emission and the attraction/repulsion of electrons floating in the space charge.

Viewed externally, the device above functions as an electrical check valve, allowing current to flow in one direction but not the other. In fact, early texts refer to it as a "vacuum valve." The term "vacuum tube" is also used, though I prefer the phrase "thermionic diode." The prefix "di" in the word "diode" is derived from the Greek word for "two." *Diode* tells us that this particular tube contains *two* elements, an electron-emitting filament, and a collector plate.

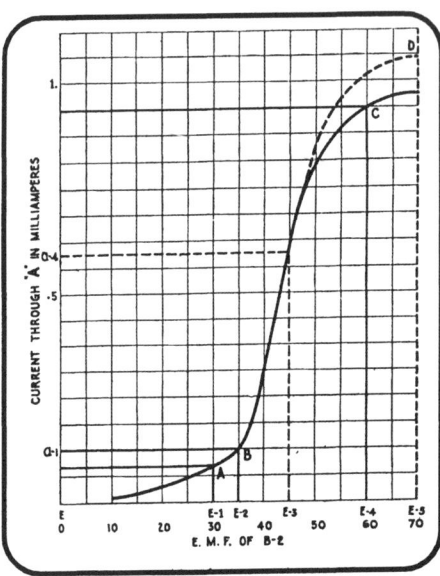

Figure 8.9
Current flowing through a thermionic diode as a function of applied voltage
(From *Vacuum Tubes in Wireless Communication*, 1918)

The flow of current through an ordinary piece of wire is uniform, linear, and directly proportional to the applied voltage. It follows Ohm's law. The greater the applied voltage, the higher the current will flow. If you actually plot this relationship on a graph, where voltage lies on the "X" axis and current lies on the "Y" axis, you end up with a very straight line that leans somewhat to the right. See figure 8.8.

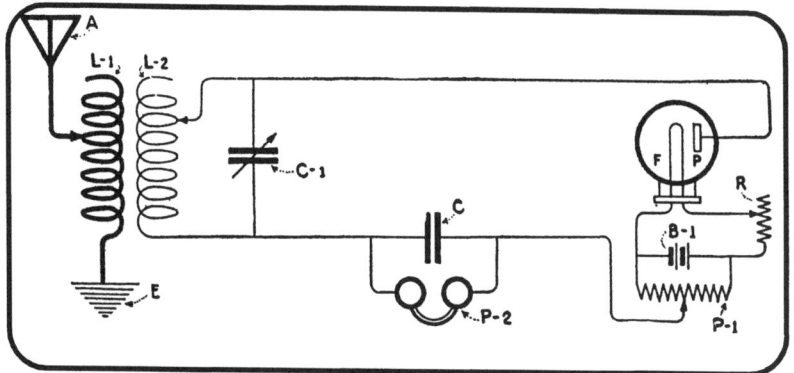

Figure 8.10
Use of the Fleming Valve as a detector in a simple radio receiver
(From *Vacuum Tubes in Wireless Communication*, 1918)

A vacuum-tube diode is not a linear device, and the flow of current through it is definitely not uniform. This becomes evident if you apply a series of ascending voltages to the "B" battery circuit and measure the flow of current in each instance. If the collected data is plotted on a graph, you'll probably wind up with the flattened S-shaped curve shown in figure 8.9. Why isn't it a straight line, and what does the shape tell us?

Starting at the leftmost end of the curve, we find that it rises very slowly, even with substantial increases in plate voltage. This is caused by the space charge. Even though the plate is positive, and attractive, electrons emitted from the filament are hindered by the repulsive effects of the cloud.

Once the positive charge on the plate gets high enough, it exerts enough attractive influence to neutralize the action of the space charge. This is depicted by the middle section of the plotted curve. As the plate voltage continues to rise, the plate becomes more and more attractive to electrons. The greater the quantities of electrons "sucked" into the plate, the more current flows through the tube. Notice that this segment of the curve is reasonably straight, rising steeply with fairly small changes in voltage. Over this limited span, the diode comes pretty close to the behavior of a an ordinary piece of wire or similar conductor.

Eventually, if the voltage is raised high enough, the plate's attractive qualities become so powerful that it can absorb electrons faster than the hot filament can release them. Once this point is reached, pushing the plate voltage even higher gets you nowhere.

There simply aren't enough free electrons available, so the graph plateaus as seen at the upper right. Under these conditions, the tube is said to be *saturated*.

Figure 8.10 is reproduced from Bucher's 1918 text, and depicts the application of a "Fleming Valve" (vacuum tube diode) to a simple radio receiver circuit.

The Thermionic Triode

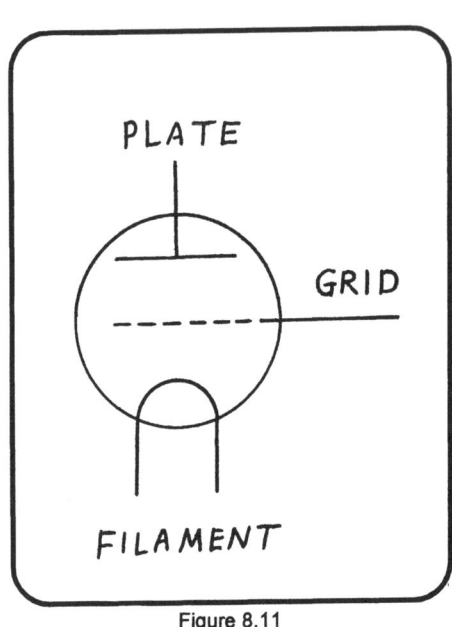

Figure 8.11
The basic triode

Lee DeForest's stroke of genius was to add a third element to the bulb, creating a thermionic *triode*. Between the filament and the plate, he installed a wire mesh or grid. See figure 8.11.

Recall that electrons are almost unbelievably microscopic in size, and infinitely tiny in comparison to the openings in the mesh. Assuming the grid isn't connected to anything, and all other factors being equal, most electrons emitted from the filament sail effortlessly through the grid and on to the plate. For all intents and purposes, the tube behaves like the diode described above.

The significance of the grid becomes evident when an electric charge is placed upon it. If a tiny negative charge is applied, electrons passing through it experience a certain degree of repulsion. (Remember, electrons carry a negative charge, and like charges tend to repel each other.) The more negative the grid becomes, the greater its ability to discourage traveling electrons from passing through it. Eventually, if the negative charge becomes great enough, the grid becomes an impenetrable barrier, and the flow of current from the filament to the plate drops to zero. Figure 8.12 depicts a simple circuit for applying various voltages to the grid while measuring the flow of current through the tube. Figure 8.13 is a graph showing the change in current through the tube as the applied grid voltage is manipulated.

Notice that there is a distinct relationship between the voltage applied to the grid and the current passing through the tube. The grid voltage actually *controls* the plate current. This is an obvious restatement of a theme that recurs throughout this book! A weak, external signal, applied to the grid, has the ability to control the flow of current supplied by a stronger, local power supply (the "B" battery). Yes, the triode is an amplifying device!

Why replace the mechanical amplifiers with thermionic triodes? There are numerous reasons. First, we've already touched upon the mechanical amplifier's undesirable sensitivity to external vibration. Admittedly the triode, with its hair-like filament and glass construction, is delicate. However, its operation is based in the manipulation of electric charges, not in the motion of mechanical parts. A vacuum tube, properly manufactured, can be extremely robust and tolerant of vibration.

Moreover, the absence of moving parts means the triode does not share the frequency limitations associated with mechanical amplifiers. The highest frequency a primitive triode can amplify is easily a million times greater than the highest frequency a carbon-based mechanical amp can reach. Triodes can be used for all sorts of applications where a mechanical amplifier never stands a chance.

We've explored a few simple receiver circuits in previous chapters, and demonstrated how various electromechanical devices

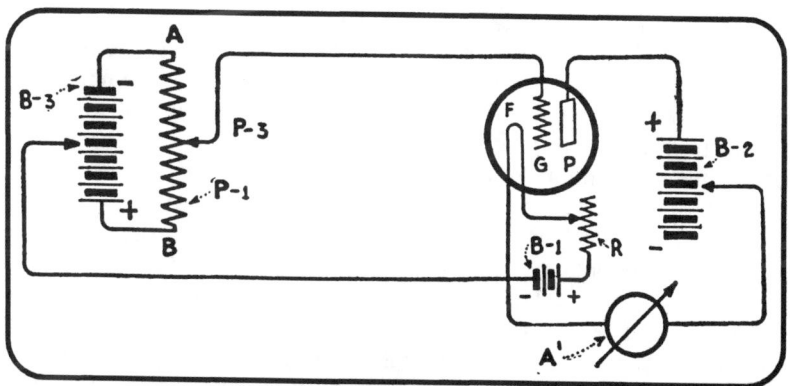

Figure 8.12
A circuit for characterizing the behavior of the triode
(From *Vacuum Tubes in Wireless Communication*, 1918)

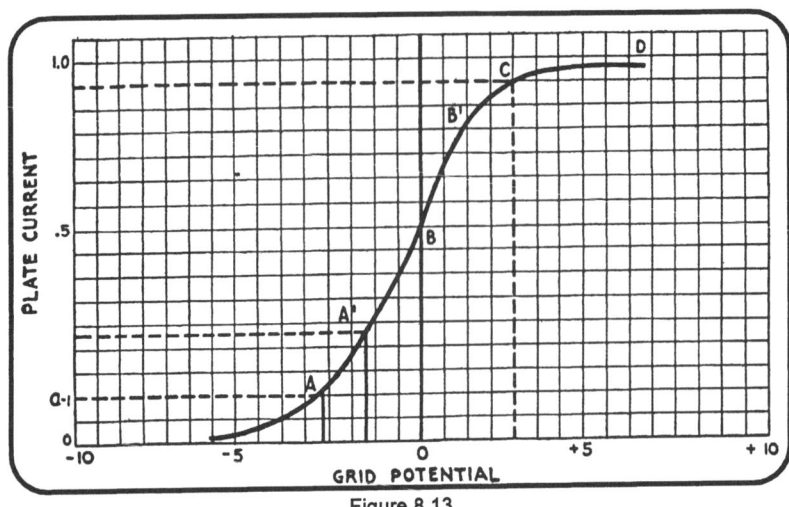

Figure 8.13
Plate current as a function of variation in applied grid voltage
(From *Vacuum Tubes in Wireless Communication*, 1918)

can be used to amplify audio signals. A triode can be made to replace any amplifying relay or sound intensifier, and perform better for the reasons just explained. If nothing else, it's much less inclined to corrupt or color the content of the audio signal.

The triode's real superiority emerges in the tuner sections of the receiver *prior* to the point where the radio signal is decoded and converted to audio. Unlike the mechanical amplifier, the triode can be used to boost the *high-frequency radio signal itself*, giving a triode-based receiver much greater sensitivity.

As we've seen, the input signal to an electromechanical amplifier is generally applied to an electromagnet of some type. The coil may be wound to be a high-impedance type, which helps to minimize the burden it places on the signal driving it. Just the same, some minimal amount of current must be drawn from the source in order to drive that coil and initiate the operation of the mechanism.

The triode has no such coil, so there's no need to draw any significant amounts of current from the source. In fact, the mere *presence* of the signal on the grid creates the electrostatic fields that regulate the flow of electrons through the tube. Thus, the input impedance of a triode can be so high that the load it places on a signal source is negligible. Vacuum tubes can be driven by extraordinarily tiny signals.

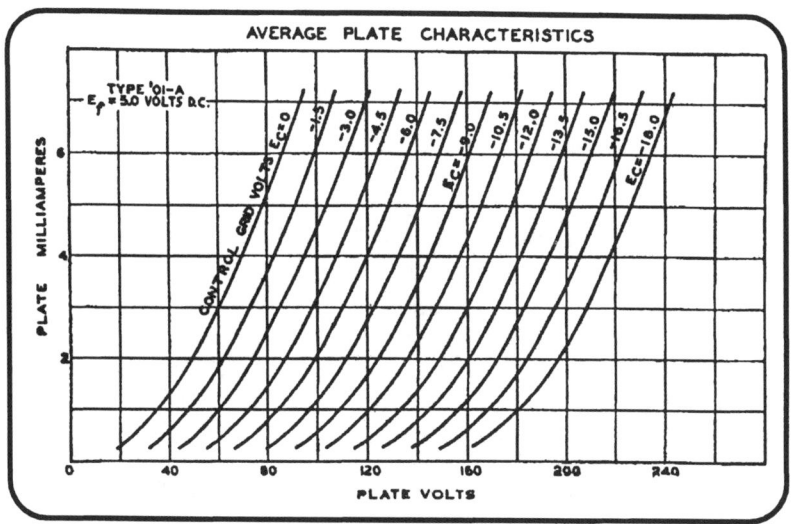

Figure 8.14
The family of curves describing the behavior of the "01A", a primitive triode

There are a number of ways to express the amplifying ability of any given triode. If you refer to previous chapters wherein we calculated the gain of our unstable contact amplifiers, you'll recall that gain is specified as the ratio of output to input. Stated another way, we measure the input in some way, in terms of voltage, for example, and then we measure the output in the same fashion. Dividing the input value into the output value tells you how many times greater the output is. This is the device's gain.

Computing gain in the triode is a bit irregular, because the tube is controlled by an input *voltage* on the grid, yet it expresses its output in terms of a change in plate *current*. Voltage and amperage (current) are certainly related through Ohm's law, but they're really apples and oranges. If you divide the former into the latter, you get a number engineers call the *mutual conductance* of the tube. Mutual conductance is generally represented by the symbol G_m. It's basically an expression of the slope of the curve in figure 8.13.

In order to compute the usual gain value, we have to compare like entities, volts to volts, or amperes to amperes. How can that be done? In the preceding pages, we've learned that there are at least two ways to affect the plate current. We can vary the plate voltage (the "B" battery supply), or we can vary the voltage applied to the grid.

Page 109

Since both actions result in a change in plate currents, we can use the ratio between them to express the tube's gain.

Physically, this amounts to varying the "B" battery voltage by an arbitrary, fixed amount, and noting the change in plate current. Next, we vary the grid voltage by an identical fixed amount. Again, we measure the change in plate current. If we take the plate current value in the first instance, and divide it by the plate current measured in the second instance, we're left with the gain value for that tube. Symbolically, triode gain is represented on paper by the Greek letter mu (μ).

Interestingly, if we divide the gain of a tube (μ) by its mutual conductance (G_m) you get a new value called R_p. Like any medium supporting the flow of electric current, the tube and its internal structures present a certain resistance to the flow of electricity. R_p is a measure of that resistance in ohms.

To this point, all of the signals we've applied to the grid were negative with respect to the filament. Can a triode handle a positive signal? The answer is yes, with certain caveats.

Negative grid voltages have the effect of restricting the flow of current through the tube, because of the repulsion that occurs between the grid and electrons emitted from the filament. The more negative the grid is, the greater a barrier it poses to electrons on their way to the plate.

A positively charged grid has an opposite effect. A positive grid will tend to attract electrons, and encourage their migration to the plate. This can actually enhance the flow of current through the tube.

The problem with positive grid signals is that a few electrons, attracted to the grid, "decide" that the extra travel to the plate is not worth their trouble. This results in the flow of electric currents in the grid itself, a condition that's not desirable in many circuits. If the grid potential is raised high enough, the stream of electrons issuing from the filament may forego the trip to the plate altogether, and elect simply to collect on the surface of the grid wires. In effect, the grid becomes the plate, and the tube has been transformed into a poor diode.

The electrical behavior of a simple diode can be represented by a single curve like the one depicted in figure 8.9. In the case of the triode, there is no one, single curve representing its behavior because it depends on what voltage is applied to the grid at the time. For this reason, triode behavior is graphically represented with a *family* of curves. Each curve in the family is labeled with the grid voltage that produced its shape. An example of such a family of curves can be seen in figure 8.14. A graph of this type speaks volumes about the behavior of a given tube, and assists engineers in properly utilizing the tube in a

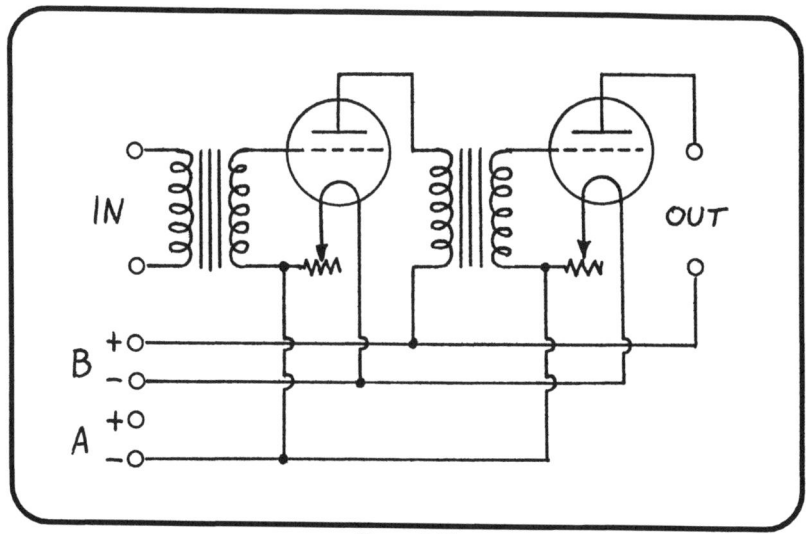

Figure 8.15
A primitive triode amplifier circuit
(Adapted from *How to Build Your Radio Receiver*, 1924)

given circuit.

Figure 8.15 is a classic design for a thermionic triode audio amplifier, circa 1924. It embodies much of what we've discussed, with a few added features.

There are two tubes used in this design. Note that they are coupled to each other and to the input signal through a set of transformers. The transformers not only provide matching to ensure that the tubes are driven and loaded properly, but they also provide the additional benefit of isolation between successive stages. This prevents unintended (direct) currents from leaking from one tube to the other and upsetting its neighbor's operation.

Note also, the variable resistors placed in the filament circuit of each tube. Variable resistors allowed the operator to adjust their respective filament intensity. In the early days, this helped prolong the life of the frail and expensive tubes and gave the operator some control of the circuit's overall gain.

Wagon Wheels and Vacuum Tubes

In an oblique way, vacuum tubes and wagon wheels are alike. Both are quaint artifacts from a distant past, and both represent technologies that have become obsolete. A cursory glance reveals both to be laughably primitive in theory, function, and construction.

Or *are* they? I challenge anyone to try building themselves a set of wagon wheels. I'm not talking about the decorative fakes found on lawn ornaments, but the *real* thing. I'm talking about a set of wheels you could slap on a wagon to haul a driver and a quarter ton of cargo a few hundred miles. Frankly, I'd be surprised if a handful of readers in all of North America could succeed. Why?

It turns out that the construction of a *real* wagon wheel takes a great deal of practical know-how. To begin with, only certain woods, like oak, hickory or elm, are suitable materials. The hubs are turned on a lathe, slotted to accept the spokes, and then fitted with four iron bands to prevent splitting. Depending upon the size of the hub, the wood used may have been seasoned for up to 10 years! Spokes are carved from seasoned oak, at least two years in age. The best spokes are given oval rather than circular cross-sections. The round portion of the wheel is composed of several curved segments called felloes. The spokes are forced into holes bored in the felloes, and the assembled wheel is fitted with an iron tire. The tire is sized to be a trifle bit too small to fit, so it's heated to a dull red, which causes it to expand. It's forced onto the wheel, which is then immersed in water. The contraction of the iron holds the entire assembly together without so much as a single nail. By the way, the spokes don't radiate perpendicularly from the hub. Instead, they're made to project from the hub at a slight angle. A properly made wheel has a noticeable "dish" to its shape, which greatly increases its durability. Then there is the matter of the internal bearing, or *box*, as it's called. The list of terms, tools, and techniques associated with wheel-making stretches on and on.

In truth, the wagon wheel may be obsolete, but it's definitely not primitive. In fact, it represents a degree of technical sophistication and advancement that only becomes evident *when you actually try to build one.* Likewise, building real vacuum tubes requires the consideration of a virtual mountain of details. Let's touch upon a few of them now.

The operation of any thermionic vacuum tube depends upon the emission of electrons from its heated filament. The hotter a given material is, the more agitated its atoms become. More electrons are emitted, so the tube is able to carrier greater currents. The catch is that high temperatures are tough on filaments. Many materials simply melt and fall to pieces before they can produce useable quantities of electrons. Carbon has a high melting point, but it tends to be mechanically fragile. Thus, materials like tungsten found use in early tube filaments. This was, in part, due to its high melting temperature, and its greater resistance to thermal shock. These same characteristics justify its use in common lightbulbs to this day.

Obviously, filaments don't light themselves. It takes energy from the filament supply to heat them to incandescence. While this is crucial to the operation of the tube, it's still considered wasteful by engineers, because it doesn't contribute directly to the amplification process. It's a burden, like a heavy canteen carried with you on a trek across the desert. The canteen is crucial to your success, but in and of itself, it contributes nothing to the goal of getting from point "a" to point "b." Hungry filaments are especially unwelcome in applications where the electronics must be powered from limited sources of energy, like batteries.

Different atomic elements have different electron configurations orbiting their nuclei. Because of this, each responds somewhat differently to thermal agitation. Some materials hold their electrons jealously, while others are more lax, and more inclined to let an electron escape here and there. One way to quantify this inclination is with a value called the *work function*. The work function of a material is a measurement of the amount of work required to set an electron free from its surface. See the table below for examples of work function values.

If it's the engineer's goal to reduce the amount of energy wasted heating a tube's filament, it serves him well to evaluate filament materials on the basis of their work function. The lower the work function, the easier it is to get a given material to emit electrons. Note that in this particular table, tungsten has the highest work function and ranks among the poorest of choices! It's a good thing that it has a high melting point and can be heated to a brilliant white, because on the basis of its work function, it's *got* to be heated aggressively in order to coax the release of electrons.

The problem with picking alternate materials from this list is that they may have other physical attributes that discourage their use as filaments. Perhaps their melting point is too low, they're too hard or brittle to be drawn into fine wire, or they're chemically volatile or reactive.

A successful strategy is to combine materials to enjoy the best attributes of both. Tungsten, for example, was sometimes combined with a small quantity of thorium oxide (1-2%), and then subjected to special heating operations that decomposed the oxide into a uniform layer of metallic thorium. This *thoriated filament* can emit as many electrons as plain tungsten, but at a lower temperature (1,700 degrees C) and at the expense of far less filament current.

Work Functions

Substance	Work Function
Tungsten	4.52
Platinum	4.40
Molybendum	4.30
Carbon	4.10
Silver	4.00
Copper	3.70
Thorium	3.40
Aluminum	3.00
Magnesium	2.70
Nickel	2.80
Titanium	2.40
Sodium	1.82
Barium	1.70
Potassium	1.55
Rubidium	1.55
Cesium	1.36

A variation on this theme was the use of platinum filaments, coated with sequential layers of barium and strontium oxides. These were baked onto the filament at high temperatures. A third technique involved the plating of a thin layer of copper to the surface of a tungsten wire, which itself was then coated with barium azyde. The azyde was decomposed with heat, resulting in the deposition of a thin layer of pure barium.

Appleton's 1931 work discusses these processes in somewhat greater detail, and notes the filament emission efficiency of each type. Compared with the plain filament, the thoriated filament offers 20 times better efficiency. The barium/strontium oxide filament measures 25 times better. The barium azyde technique reigns supreme with a seventy-fold improvement over the plain tungsten equivalent!

A more recent reference in the *RCA Receiving Tube Manual* (1971) describes filaments composed of nickel-alloy wires or ribbons, heavily coated with "alkaline earths." These materials are so effective, the filament needs no more heat than that required to bring it to a dull red glow (700-750 degrees C).

The use of a filament as the electron source in a vacuum tube presupposes that a battery or other direct current source is used to light it. Direct current results in steady, uniform heating, accompanied by the steady, uniform emission of electrons.

Alternating current, like the 60-cycle power found in your wall outlet, doesn't work with filaments as well. Filaments are, by nature, thin and lightweight, so they have little mass and little ability to store heat. Alternating currents fluctuate, which results in minor fluctuations in filament temperature. In the end, the electron emission varies, which permeates and corrupts any amplified signal with an annoying hum.

Still, a vacuum tube that can be heated with alternating current is a desirable device, because it's suitable for use in electronics that will be powered from a wall socket. As luck would have it, the solution is simple.

If you think about it, a filament actually performs *two* functions. First, it heats itself, and at the same time, it acts as the electrode from which electrons are emitted. In more advanced tubes, these functions are separated through the addition of an electrode called a *cathode*. In these tubes, the cathode becomes the source from which thermionic electrons are emitted. It may or may not be coated with other substances in order to reduce its work function and therefore improve its emission efficiency.

The cathode is brought to operating temperature with a special filament called a *heater*. The heater filament is not designed to emit electrons. It serves no purpose other than to heat the cathode, and is electrically insulated from it.

Cathodes are designed with greater physical bulk than the threadlike filament wires. They have the capacity to retain heat. They've got sufficient thermal mass to hold a constant temperature, even if their heaters are being driven with alternating currents.

When cathodes are added to vacuum tube diodes or triodes, the heaters aren't figured into the functional electrode count. A rectifying tube with a heater, cathode, and plate is still considered a two-element diode. An amplifying tube containing a heater, cathode, grid, and plate is considered a three-element tube, or triode, because the heater serves no electrical purpose, it merely facilitates the operation of the cathode.

Over the years, control grids have taken a variety of forms. Composed of molybdenum or nickel (1930's reference), grids began as simple wire zigzags, evolving to flat grilles, disks or squares of wire cloth mesh, and eventually cylinders composed of rolled mesh or wire wound in loose spirals. In this latter design, the electrodes are coaxial, i.e., the filament lies at the center surrounded by the grid, which is in turn surrounded by a cylindrical metal plate.

Since the grid is the element that "throttles" the current passing from filament to plate, it stands to reason that many of a given triode's behavioral characteristics are linked to the geometry and placement of its grid.

For instance, the spacing between adjacent wires in the grid (which determines the size of gaps through which electrons may pass) impacts the tube's gain. A tight spacing produces tubes with greater gain, but less ability to handle large signals.

There is no rule that says the grid's density must be uniform. In fact, there is a whole class of tubes called *remote-cutoff* tubes, which feature a control grid with tight windings at its ends and relatively open spacing near the middle. The gain in a tube of this type actually varies depending upon the operating conditions. Remote-cutoff tubes tolerate large input signals better than tubes with uniform grids.

The internal resistance of the tube is inversely proportional to the area of the cathode and plate. Roughly speaking, the larger the plate, the lower the plate resistance. Tubes, like other amplifying instruments, tend to work best when driving a load that matches their internal resistance. Matching transformers, like the one described several chapters back, are selected to conform to the electrical characteristics of the plate.

The spacing of each electrode relative to the others affects the operation of the tube as well. According to one source, amplifier tubes work best when the grid is placed as close to the plate as possible, and as far from the filament as practical.

On the other hand, if you've read *The Voice of the Crystal* or similar texts, you'll recognize that the grid and plate, in close proximity, have all the necessary qualifications to act as a capacitor (or condenser, as they used to be called).

A "hidden" capacitor linking the grid to the plate is a bad thing, because it provides a high-frequency leakage path from the output of the tube (the plate) back to the input (the grid). If you've ever witnessed a stage performer or lecturer carry his microphone too close to the speakers, you're already familiar with the process of *feedback*. Ambient noise enters the microphone where it is amplified and fed to the speakers. The sound from the speakers re-enters the microphone, is amplified once more, and sent back to the speakers. Again, and again, the sound energy returns to the microphone, looping furiously through the system and gaining energy at each iteration. The result is the annoying shriek or squeal that causes the

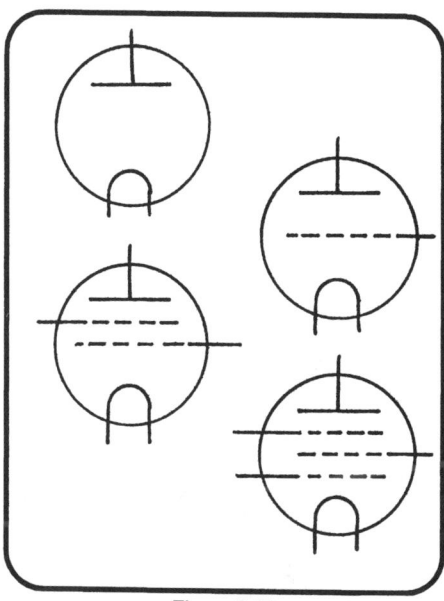

Figure 8.16
Variations in vacuum tubes: the diode (two electrodes), triode (three electrodes), tetrode (four electrodes), and pentode (5 electrodes)

audience to clench their teeth and plug their ears. A similar effect can be found with triodes, though the "shriek" produced may be so high a frequency as to be inaudible.

Inter-electrode capacitance can be reduced several ways. One way is to reduce the area of grid and plate. Another is to simply space them farther apart. However, in the context of the previous paragraphs, these steps may be objectionable. The tetrode, or four-element tube, was developed to address the capacitance problem.

A tetrode is similar in construction to a triode, but it features an additional grid, called a *screen grid*, which is inserted between the control grid and the plate. With a proper positive charge, it acts as shield, which can reduce the capacitance between the two by a factor of 100 or more. This reduction in the undesirable coupling between output and input means the tube can be operated at much higher frequencies than an otherwise identical triode.

Electrons emitted from the a tube's cathode invariably end their journey with an impact on the surface of the plate. Interestingly, if an electron is particularly energetic, it may strike the plate hard enough to knock other electrons out of the plate material. The emission of an electron resulting from the impact of another electron is called *secondary emission.*

Secondary emission is usually not a problem for triodes. Any electrons knocked off the plate tend to linger around it, because the plate has a positive charge. Eventually, they return to it. The situation is more complicated with tetrodes, because of the positively charged screen grid. Secondary electrons may find themselves attracted to the screen grid. This leakage of electrons away from the plate is undesirable and causes peculiarities in the amplifying characteristics of the tube.

The anomalies found in the tetrode are corrected with the addition of yet another grid, called the suppressor grid. The suppressor grid is inserted between the screen grid and the plate, and it's typically given a negative charge. This tube configuration contains five elements, the cathode, control grid, screen grid, suppressor grid, and plate. It's referred to as a pentode. The pentode solves the problem of secondary emission, because any electrons driven off the face of the plate will encounter the suppressor grid, whose negative charge will repel the fugitive electrons back to the plate where they belong. See figure 8.16.

References

Appleton, E. V., *Thermionic Vacuum Tubes*, copyright 1931, E. P. Dutton And Company, New York, pp. 5-14, (construction of vacuum tubes)

Ballard, William C., *Elements of Radio Telephony*, copyright 1922, McGraw-Hill Book Company, Inc., pp. 14-16, (historic knowledge of the peculiar properties of heated metals)

Banning, Kendall, and Cockaday, L. M., *How to Build Your Radio Receiver*, copyright 1924, Popular Radio Inc, New York, p. 39, (audio frequency amplifier circuit)

Bucher, Elmer E., *Vacuum Tubes in Wireless Communication*, copyright 1918, Wireless Press, New York, pp. 26-28, (space charges in the diode, saturation, and electrode spacing)

Chaffee, E. Leon, *Theory of Thermionic Vacuum Tubes*, copyright 1933, McGraw-Hill Book Company, Inc., New York and London, pp. 5-16, (historical development of the vacuum tube), pp. 114-115, (monatomic films, and production of thoriated filaments)

Condon, E.U., and Odishaw, Hugh, Editors, *Handbook of Physics*, copyright 1967, McGraw-Hill Book Company, New York, section 9 dash 4, (ratio of nuclear diameter to atomic diameter)

Francis, Charlotte A., and Morse, Edna C., *Fundamentals of Chemistry and Applications*, copyright 1958, The MacMillan Company, New York, pp. 117-123, (basic atomic structure)

Ghirardi, Alfred A., *Radio Physics Course,* copyright 1933, Radio & Technical Publishing Company, New York City, New York, p. 224 (information on inductive reactance), pp. 384-385, (the Edison Effect and electronic emission from solids)

Henney, Keith, *The Radio Engineering Handbook*, copyright 1941, McGraw-Hill Book Company, New York, p. 237, (work function of substances), pp. 238-239 (space charges), p. 247, (amplification factor), p.248, (equations depicting the effect of electrode geometry to gain)

Lange, Norbert Adolph, PH. D., *Handbook of Chemistry*, copyright 1956, Handbook Publishers, Inc., Sandusky, OH, p. 100, (physical constants of the elements), p. 1696, (mean free path of particles are varying atmospheric pressures)

Martin, L.H., and Hill, R.D., *A Manual of Vacuum Practice*, copyright 1947, Melbourne University Press, reprinted by Lindsay Publications, copyright 1997, ISBN 1-55918-188-5, p. 113, (mean free path)

Moyer, James A., and Wostrel, John F., *Radio Receiving Tubes*, copyright 1931, The Maple Press Company, York, PA, pp. 1-9, (history of the vacuum tube), pp. 9-23, (construction of radio tubes), pp. 56-58 (space charges and characteristic curves), pp.108-109 (electrical characteristics for common tubes)

RCA Corporation, *RCA Receiving Tube Manual*, copyright 1971, RCA Corporation, Harrison, NJ, pp. 3-12, (internal construction of vacuum tubes), pp. 13-14, (electron tube characteristics)

Selman, M.D., Joseph, *The Fundamentals of X-Ray And Radium Physics*, copyright 1972, Charles C. Thomas, Publisher, Springfield, IL, pp. 36-53, (the structure of matter)

Thompson, Silvanus P., *Electricity and Magnetism*, copyright 1905? Thompson and Thomas, Chicago, pp. 11-16, (electroscopes)

Van Der Bijl, *The Thermionic Vacuum Tube and Its Applications*, copyright 1920, McGraw-Hill Book Company, New York, pp. xii, (O.W. Richardson, and his study of kinetic activity in electrons, and explanation of thermionic emission)

Wehr, M. Russel and Richards, Jr., James A., *Physics of the Atom*, copyright 1960, Addison-Wesley Publishing Company, Inc., London, England, pp. 78-110, (atomic models of Rutherford and Bohr), pp. 107-108, (present disposition of the Bohr model)

Zworykin, V. K. and Wilson, E. D., *Photocells and Their Application*, copyright 1930, John Wiley & Sons, Inc., New York, p. 33, (least work functions), pp. 49-60, (methods for the deposition of alkali metals)

http://www.scri.fsu.edu/~capstick/AST1002/Chapter6/Starlight.htm

Capstick, Simon, "Atoms and Starlight," copyright unknown, Florida State University, Department of Physics, Fall, 2000, (relative size of nucleus to first electron orbit)

Chapter 9
Vacuum Basics

A s we've seen in the previous chapter, the process of thermionic emission is most effective when the heated filament or cathode lies inside of an evacuated container. The more complete the vacuum, the fewer residual particles remain to interfere with the movement of electrons. Before we attempt to fabricate our own thermionic amplifying devices, it's worth our while to understand what terms like *pressure* and *vacuum* really mean, and how they can be quantified.

Atoms and molecules are in constant, vibratory motion. Though this may not be apparent in solid materials where the constituent atoms are bound tightly together, it's quite evident in fluids. It's especially true in the case of gases, whose molecules exhibit extreme mobility.

Let's consider a rigid, sealed vessel filled with some quantity of gas. For the sake of argument, we'll presume the gas is nitrogen. Nitrogen gas is made up of binary molecules, each of which contains two nitrogen atoms. These molecules float about, dancing and jiggling in response to the heat energy present in all matter. (The hotter a substance is, the more energetic its molecules. *Absolute zero* is the temperature at which all vibratory motion ceases.) As these molecules zing to and fro within the confines of the container, they eventually collide with, and bounce off its walls.

The effect of molecular collisions may be easily envisioned if we take a small cardboard box and toss in a handful of tiny pebbles. If the box is closed and vigorously shaken, the rattling pebbles can be thought to represent gas molecules in motion. Note that although each pebble is lightweight and seemingly insignificant, their collective influence on the walls of the box become measurable and noteworthy.

The impact of gas molecules on the walls of a container is essentially random, and uniformly distributed over the vessel's internal area. If we take a measurement of the total collision force, and divide that by the total area of the interior of the container, we arrive at a means to quantify the gas pressure in terms of force per unit area. A common unit of pressure is the pound per square inch, or *psi*.

Earth is blanketed by an atmosphere whose pressure at sea level amounts to just under 15 psi. My desktop measures 30 x 60 inches, for a total surface area of 1,800 square inches. If you multiply that area by atmospheric pressure, you'll discover that my desk somehow supports an astonishing 27,000 pounds!

Ridiculous, you say! That desk should be crushed flat! Not really. Air is a fluid, and it flows to all sides of the desk. The desk withstands the 27,000-pound force on its top surface because the air exerts the same force to the bottom side of the desk. The desk is unscathed, because the various forces are balanced, and simply cancel out.

One of the most impressive demonstrations of air pressure was conducted by Otto von Guericke in 1654. Von Guericke fashioned two copper hemispheres, each 22 inches in diameter. The open faces of the two domes were brought together, joined only by a gasket. The gasket was leather, ring-shaped, and had been thoroughly soaked in wax and oil to help effect an airtight seal. Von Guericke attached a primitive air pump, and removed as much air from the interior of the resulting sphere as possible.

To the amazement of Emperor Ferdinand III, who was in attendance, two teams of eight horses, chained to opposite ends of the sphere, were unable to yank the halves apart! In truth, the horses never had a fighting chance, as the force necessary to separate the domes amounts to nearly three tons!

A casual observer may be inclined to conclude that the sphere was held together by the suction inside of the sphere. This is absolutely *incorrect*. By exhausting the air from the interior of the sphere, Von Guericke simply nullified the forces acting to separate the hemispheres. This left only external air pressure, whose inclination it was to keep the halves driven firmly together.

A more contemporary example of atmospheric pressure can be observed any time a drinking straw is used. Once again, an uninformed observer would attribute the rise of fluid in the shaft of the straw to the "suction" created by the drinker's mouth. Wrong!

Under ordinary conditions, the force of air pressure on the open end of the straw is equal to the force of the atmosphere on the surface of the beverage. The system is at equilibrium, so there's nothing to motivate the liquid to move. The level of the fluid inside of the straw matches the level of the surrounding liquid in the glass.

However, if the drinker reduces the pressure at the top of the straw with his lips and tongue, this creates an imbalance, because the forces applied to the liquid from the top end of the straw will now be less than the force applied to the beverage by the atmosphere. The result is that air pressure drives the liquid up the straw. Let me restate this: The beverage is not "sucked" up the straw by a vacuum, it's *driven* up the straw by a difference in *pressure*.

What, then, is a vacuum? What does the word mean? The truth is, it has no definite meaning. Rather, it depends upon your point of reference.

For example, what temperature is cold? There's no way of knowing. Like the word "vacuum," "cold" is meaningless unless you understand the point of reference. Ice is cold compared to body temperature. Boiling water, which is arguably "hot" to humans, is nonetheless quite cold compared with molten iron.

If normal atmospheric pressure is your point of reference, then one might say that a vacuum exists in the mouth of the straw's user. Most of the time, a vacuum is simply a pressure that's less than normal atmospheric pressure.

If it were possible to eliminate all particles from a fixed region of space, there could be no molecular collisions, no forces on the internal walls of the container, with a resulting pressure of zero. A region of space with zero pressure is considered to contain an *absolute vacuum*. Absolute vacuums are both interesting and useful for a variety of reasons, though they've never been achieved on Earth. Even in the deepest reaches of outer space, a few particles can be found here and there, so strictly speaking, the vacuum is not absolute out there, either. However, modern pumping equipment comes close enough for practical purposes.

Measuring Pressure and Vacuum

How is pressure measured? There are dozens of ways, each with advantages and disadvantages. For the sake of simplicity, we can classify these techniques under one of three categories, that is, mechanical, thermal, and ionization. Let's take a look mechanical measuring devices first.

As we said earlier, gases under pressure exert a specific force over the walls of a container. Knowing this, it's relatively simple to fashion a gizmo that'll respond to that force. Figure 9.1 shows a springy set of bellows linked to a pointer with a rack-and-pinion gear. If gas, under pressure, is fed into the bellows, it expands, driving the rack upward. This causes the pinion to rotate clockwise, and the pointer rises to indicate the raised pressure in the bellows.

If the bellows are connected to a suitable pump and evacuated, they collapse under the force imposed by the surrounding atmosphere. The pointer needle will rotate counterclockwise.

What if the bellows is left open to the air? In this case, the force inside the bellows will match the force outside of the bellows. The contraption will return to its rest position.

Notice that this instrument mechanically compares the pressure inside the bellows to ambient air pressure. A pressure measurement referenced to atmospheric pressure is called *gauge pressure*. Tire gauges, for example, measure gauge pressure. If an automobile tire measures 32 psi, this really means the tire pressure is 32 psi with respect to atmospheric pressure. Put another way, the tire pressure is 32 psi greater than the current atmospheric pressure, the implied point of reference.

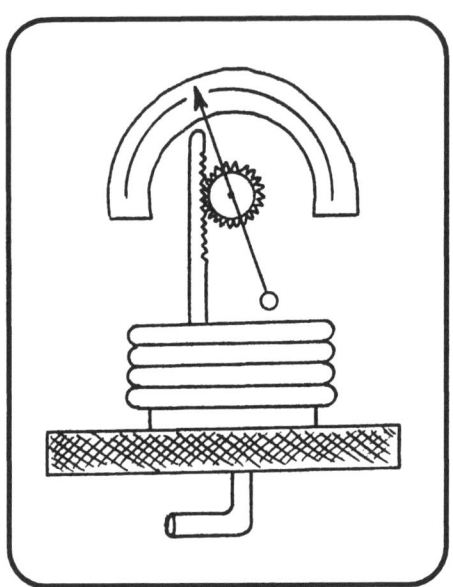

Figure 9.1
A "gauge" aneroid pressure indicator, where measurements are referenced to ambient pressure

Gauge pressure instruments are useful for lots of applications, though they're less attractive if any degree of precision is required. Remember that atmospheric pressure varies from hour to hour in response to weather conditions. If atmospheric pressure varies, your reference varies, and so will your measurement.

One way around the problem is to use a reference that's fixed and unchanging. An absolute vacuum (or as close as can be achieved) is ideal. Figure 9.2 shows a modified bellows instrument enclosed in a glass jar. The jar is evacuated, so the action of the bellows is referenced against an absolute vacuum. The instrument now reads in terms of *absolute pressure*. If the instrument's input is left dangling in open air, the dial will read something on the order of 15 psi. The automobile tire, mentioned above, will measure about 47 psi.

It doesn't take a lot of imagination to create variations on this theme. For example, the bellows can be replaced with a diaphragm that deflects in response to changes in pressure. The diaphragm can drive a pointer needle through levers, a tiny chain, fibers, cams, or whatnot.

Another interesting approach is to fabricate a delicate coil of fine glass tubing. One end of the coil is sealed, the other end is connected to the pressure source of unknown magnitude. Forces on interior walls of the tube, exerted by the applied pressure, induce a stress on the coil causing it to unwind ever so slightly. When the pressure is relieved, the coil snaps back to its original shape. The rotary motion of the coil as it flexes can be

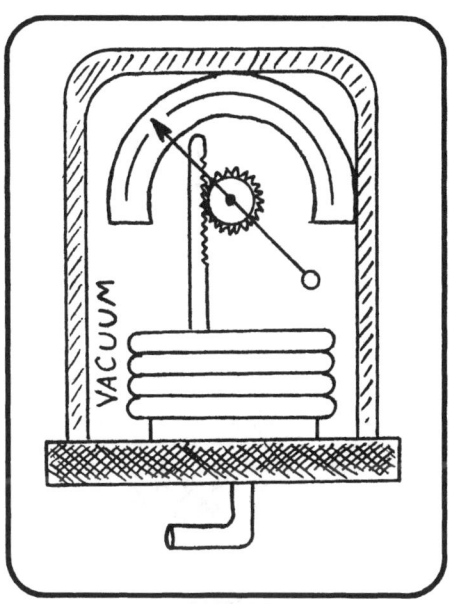

Figure 9.2
An "absolute" aneroid pressure indicator, where measurements are referenced to a vacuum

used to drive a physical pointer, or in some instances, a tiny mirror. In the latter case, a beam of light is bounced off the mirror and projected onto a printed scale. A dot of light falls upon the scale, indicating the measured pressure. Light beams make excellent pointers, because they weigh nothing and represent no physical burden to the mechanism. At the same time, they can be made very long, taking advantage of lever action to reveal the tiniest fraction of a degree of motion.

A few paragraphs back, I cited a common drinking straw as an example of a pressure working against a vacuum. Interestingly, the behavior of a fluid inside a glass tube is the basis for yet another measuring device called a manometer.

In 1643, Evangelista Torricelli sealed one end of a glass tube and then filled it with mercury. Placing his finger over the open end of the tube, he inverted it, and submerged the open end in a small dish of mercury. When he removed his finger, something strange happened. Instead of draining into the dish, the mercury level fell a small distance, but remained suspended within the tube. See figure 9.3.

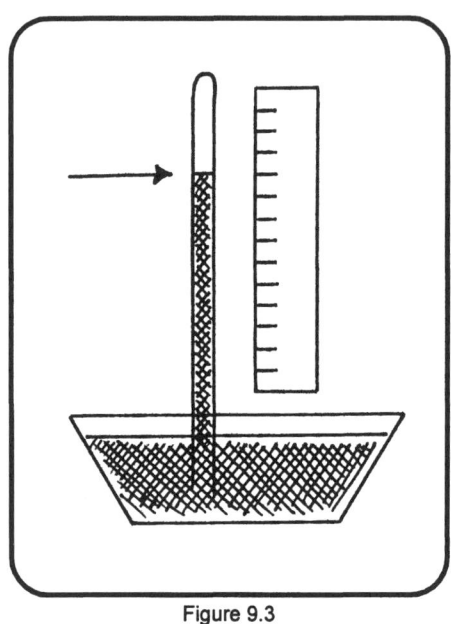

Figure 9.3
Torricelli's mercury manometer

The effect is quite explainable on the basis of what we know today. Mercury is a dense liquid. As it falls, it creates a pocket of nothingness at the sealed end of the tube. This pocket is extremely low in pressure- a vacuum, more or less. Within the interior of the tube, there's no pressure left to drive the mercury out.

On the outside, air pressure bears down on the mercury in the dish, and tends to drive it back up into the open end of the tube. It can do this to a certain degree, but it must overcome the weight of the mercury column itself. In operation, the column will fall until the external pressure applied to the manometer precisely balances the weight of the mercury column suspended in the tube. Any variation in the air pressure affects the balance of the system, so the mercury column will rise or fall in order to reestablish equilibrium. Because this relationship is specific and definite, it's possible to express pressure in terms of the height of the mercury column that it can support. All you need to do is fix a ruler alongside the glass tube.

Inches of mercury is a common unit for air pressure, and one you've probably heard used by various weather services. In fact, it was

Torricelli who first noticed that the mercury column in his instrument magically rose and fell in response to changes in the local weather.

Air pressure at sea level can support about 30 inches of mercury. If a metric ruler is placed along the side the glass tube instead, one can measure the pressure in *millimeters of mercury*; 30 inches of mercury translates into about 760 millimeters of mercury. Incidentally, Torricelli's name eventually became a unit of measure, the *torr*. Gas under the pressure of 1 torr is capable of supporting a mercury column that's 1 millimeter in height.

Notice that the manometer mechanically compares the applied pressure to a reference composed of the vacuum at the head of the mercury column. Thus, the manometer provides readings of absolute pressure.

A variation on the manometer idea is the instrument shown in figure 9.4. This is still a manometer, but the open dish of mercury has been replaced by a "U"-shaped reservoir. The source of the pressure you wish to measure is simply connected to the free end of the U. You may have seen an instrument of this type in your doctor's office. Manometers were once used with inflatable cuffs to measure blood pressure.

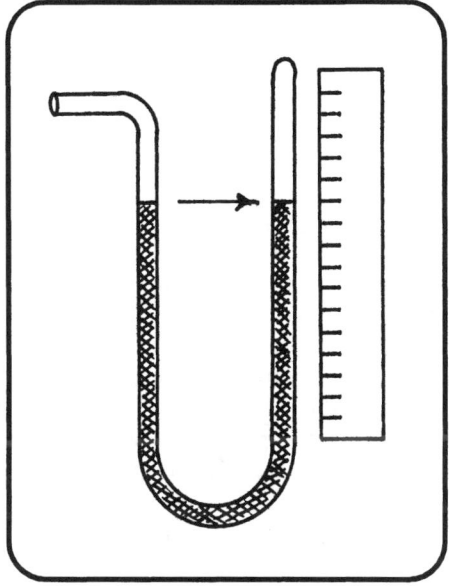

Figure 9.4
Another form of the mercury manometer

Mechanical instruments can work well at pressures well above atmospheric, and are reasonably accurate under moderate or "soft" vacuums. However, the more extreme the vacuum becomes, the fewer molecules are left to apply force to the moving parts of the gauge. If the level of vacuum descends much below a single torr, the usefulness of mechanical measuring instruments begins to dwindle. Below this level, thermal pressure gauges become attractive.

Everyone knows that a hot apple pie, fresh from the oven, will quickly cool if placed on a windowsill. Part of the cooling process involves the transfer of heat from the warm pie to the surrounding air. As air molecules collide with, and bound off of the pie, each carries away a tiny quantity of heat.

The rate at which heat can be dispersed depends on the pressure of the surrounding air. The higher the pressure, the more molecules are available to act as couriers. If the air pressure is reduced, fewer molecules are left to carry away the pie's heat.

A thermal device called a *Pirani gauge* exploits this relationship to infer the pressure of a gas. The heart of the Pirani is a segment of wire, which is heated by a precise and regulated electrical current. The wire is immersed in the gas we wish to measure. Like the hot apple pie, it transfers some of its heat into the surrounding gas. The wire cools at a known rate, dependant upon the pressure of that gas

All wire has a certain degree of electrical resistance. This resistance changes in response to temperature. Measuring the resistance of the heated wire in the Pirani gauge gives us an idea of its temperature, from which we may ultimately deduce the pressure of the surrounding gas. The technique is admittedly roundabout, but is actually simple to implement and reasonably accurate. With it, we can measure vacuums down to 0.01 millitorr or even less. (A millitorr is one one-thousandth of a torr.)

A properly equipped lab can reduce the pressure in a vessel to absurd extremes, that is, values measured in millionths or even billionths of a torr. In terms of measurement, "hard" vacuums like these are well below the useful range of thermal instruments. How does one make pressure measurements at these levels?

In the previous chapter, we discussed the manner in which air molecules can impede and degrade the flow of electrons from a heated filament to a charged collector plate. In fact, part of the reason for operating a thermionic diode or triode in a vacuum is to minimize such interference. As it turns out, there is a specific relationship between low gas pressures and the ease with which a stream of electrons can travel through it.

A hard vacuum within a sealed chamber can be measured by inserting a filament, grid, and plate into that space. With the filament lit and appropriate charges applied to the grid and plate, the magnitude of the electron current flowing from the filament to the plate will depend upon the pressure in the chamber. By measuring the plate current, the chamber pressure can be calculated.

Producing a Vacuum

A pump is a mechanical contrivance used to transfer a fluid from one place to another. While the basic purpose of all pumps is the same, as we shall see, some are better suited for creating vacuums than others.

Figure 9.5 illustrates the workings of a simple piston pump of the type one might use to compress air. It consists of a piston driven by a crankshaft that has been fitted to the interior of a cylinder. As the crankshaft turns, the piston moves up and down.

At the head of the cylinder are two ports, each fitted with a one-way valve. They're depicted here as small, spring-loaded cones, which when at rest, plug their respective ports. Notice that one points into the cylinder, and one points out.

During the piston's downstroke, the pressure within the cylinder is reduced, creating an imbalance. Air, under atmospheric pressure, drives the input (left hand) valve open, forcing the cone off its seat. Air rushes in to reestablish equilibrium, and continues to flow until the piston reaches the bottom of its stroke. The input valve then closes under the action of its spring.

When the piston is driven upward, the gas in the cylinder is compressed. Its pressure rises, so it seeks an opportunity for escape. The input valve is not an option, as it's closed, and it's orientated in such a manner that it can't be forced open from within the cylinder. However, the output (right hand) valve is set up in reverse. Gas, under pressure from the rising piston, can force the cone off of its seat. The pressurized gas can exit through the output valve and into the reservoir. Note that the one-way nature of the output valve also prevents gas from leaking from the reservoir back into the cylinder.

A pump of this type is best known as a *compressor*, as its intended purpose is to compress ambient air for filling tires, powering air tools, or spraying paint. It's logical to presume that the same pump, used in reverse, might be suitable for creating a vacuum in a sealed chamber or container. In the latter case, all you'd have to do is connect the container to the input side of the pump and "suck" the air out of it...or can you?

In practice, the performance of the piston-compressor-in-reverse configuration is likely to be disappointing. To begin with, the compressor's input valve is intended to open under the influence of *atmospheric* pressure. Under conditions of reduced pressure, there's nothing to force the cone off of its seat. Beyond a certain level of vacuum, the valve no longer opens, so nothing can enter the cylinder.

Even if it did, the same sort of problem plagues the output valve. Despite the eventual rise of the piston, the rarified gas in the cylinder can't exert enough force to pop the output valve open. If nothing else, it remains tightly sealed under the force of ambient air trying to get back into the cylinder. At this point, the piston may flail dutifully, but no additional pumping can occur.

Diaphragm compressors are cousins to the piston compressor and conceptually similar in operation, with the exception that a crankshaft-driven diaphragm replaces the sliding piston. I have knowledge of at least one commercial application where diaphragm compressors are routinely used, in reverse, as vacuum pumps. Employed in this manner (an application that's inconsistent with their intended purpose) they struggle to achieve vacuums below 10 torr. They also suffer accelerated wear and increased incidents of breakdown.

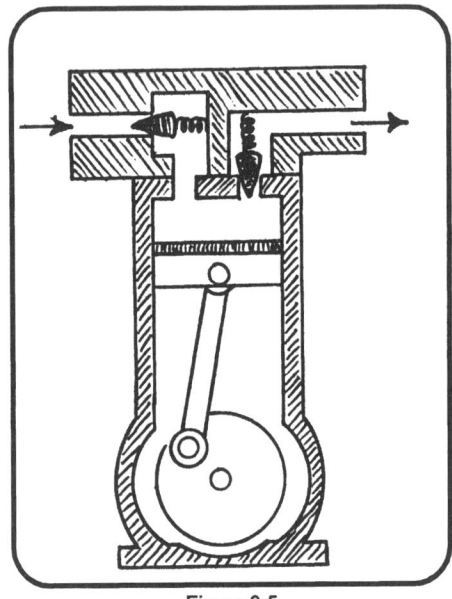

Figure 9.5
A piston-type vacuum pump

Without doubt, the most common type of mechanical pump for vacuum applications is the rotary oil pump. These are produced by a number of manufacturers in a variety of specific configurations, but figure 9.6 illustrates their basic operation.

The heart of the pump is a cylindrical pumping chamber in which turns an off-center rotor. The rotor is machined to accept two spring-loaded vanes. As the rotor spins, the vanes slide along the walls of the pumping chamber, capturing a quantity of gas from the input side of the pump, compressing it, and forcing it through a one-way exhaust valve.

A reservoir located above the exhaust port is filled with pump oil. The oil is free to trickle into the interior of the pump, where it coats all of the surfaces. In addition to reducing friction and wear, the oil acts to improve the seal between the sliding parts of the pump. Two-stage versions of the rotary pump, where two pumping chambers are connected in series, can easily achieve vacuums down to 0.5 millitorr, or even less.

As long as there are enough gas molecules present in the system, the sweeping action of the oil pump's internal parts results in a meaningful transfer of molecules from one rotation to the next. However, as the vacuum drops to fractions of a millitorr and below, air behaves less like a fluid than a smattering of independent, rambling particles that may or may not wander into the pump.

Around 1916, Irving Langmuir developed a so-called "diffusion" pump that significantly extends our ability to reach low pressures. See figure 9.7.

Note that the diffusion pump contains no moving parts. Rather, it depends upon a pumping medium, originally mercury, which was vaporized by an electric heater. The vapor is directed through guides or baffles to a point where it's developed into a jet or stream. As the vapor stream plays its way across the interior of the pump, it gathers and sweeps up gas molecules in much the same way that a stream of water from the nozzle of a garden hose can be used to whisk away sand from a dirty sidewalk.

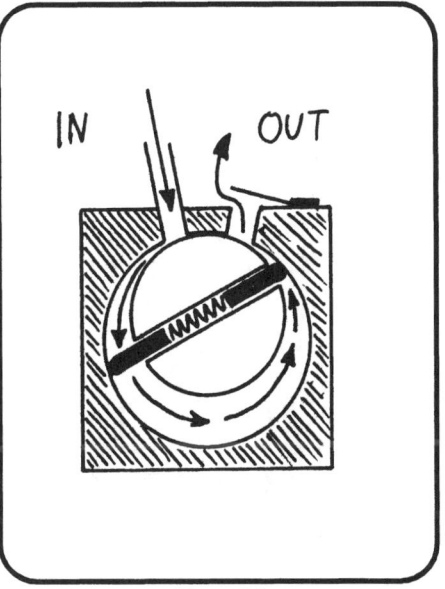

Figure 9.6
A rotary-vane vacuum pump

The diffusion pump is not used by itself. In fact, it won't operate properly at ambient pressures. Rather, it's used in conjunction with an oil vane or similar mechanical pump. The mechanical pump must first reduce the system pressure to a few torr or less. Then, with the mechanical pump still in operation, the diffusion pump can be activated.

Errant gas molecules, gathered by the mercury vapor, are swept toward the diffusion pump's exhaust port where the rotary pump can carry them away. The mercury vapor, by the way, eventually condenses and drains back into the base of the diffusion pump. There, it's vaporized again, and the cycle repeats.

Mercury is a highly effective medium for use in Langmuir's system. Unfortunately, its vapors are notoriously toxic. Mercury atoms are also liable to sneak their way out of the diffusion pump and into the system you're trying to evacuate. Cold traps can be used to prevent contamination of the item being evacuated, but many users have simply opted to employ other pumping fluids. There are a variety of special oils, for example, that can be used in place of mercury, which will deliver similar performance without mercury's drawbacks.

How well does the diffusion pump work? *A Manual of Vacuum Practice* references a particular oil-based diffusion pump with a lower pressure limit at 5 x 10^{-8} torr. (This is 0.00005 millitorr!)

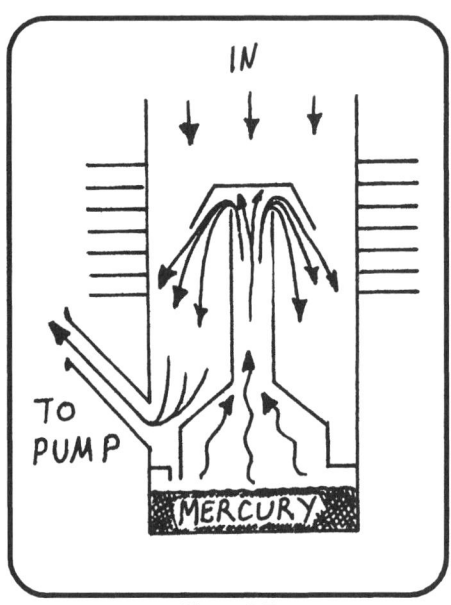

Figure 9.7
Principle behind the mercury-diffusion pump

Thermionic Tubes and the Challenge of Working With Vacuums

Traditionally, vacuum tubes were fashioned primarily from glass. In the 1930's, the construction of a vacuum tube began with the fabrication of a glass electrode support or stem. Embedded in the glass were assorted segments of wire intended to support the tube's internal components, and to provide electrical connection to the outside world. The stem was also fitted with a small, glass pipe, through which the air would later be withdrawn from the bulb.

Next, the filament, grid(s), and plate were welded or otherwise attached to the electrode wires. The completed assembly was inserted into a glass bulb, which had been prepared earlier. The stem and the mouth of the bulb were fused together, forming an air-tight seal. At this stage, the vacuum tube itself was complete.

Of course, there was the matter of evacuation. Vacuum pumps were attached to the glass pipe protruding from the stem, and as much air was withdrawn from the interior of the tube as was possible. References from the 1930's cite pressures in a range from 1 millitorr (*Thermionic Vacuum Tubes*) to roughly 0.005 millitorr (*Radio Receiving Tubes*). A pressure reference in the *RCA Receiving Tube Manual* works out to about .008 millitorr.

One would think that sealing off the evacuation pipe with a torch would mark the end of the story. In reality, it's quite difficult to maintain a vacuum, even in a supposedly "sealed" container! Let's look at some of the pitfalls.

Glass, like most materials, expands and contracts in response to changes in temperature. This is also true of metals, like those that might be used for electrode wires. Different materials react to temperature changes to varying degrees. This is a real problem if two materials must remain in intimate contact in order to assure an air tight seal.

If a wire, passing through the glass wall of a vacuum tube, expands or contracts at a rate other than the rate associated with the glass, two outcomes are possible. The metal may shrink away from the glass, allowing the leakage of air into the bulb, or the metal may swell and induce stress in the glass. Stress ultimately leads to fractures, with the resulting loss of vacuum in the bulb. Special formulations of glass, and specific metal alloys can be chosen to match, as closely as possible, their respective rates of expansion.

Most people are aware that, at sea level, water boils at 100 degrees C. At higher altitudes, where the ambient air pressure is less, the boiling temperature is less. It's possible, in fact, to place a bowl of water in a sealed chamber, draw a vacuum, and reach a point where the water in the bowl will actually boil at room temperature.

It stands to reason that the same holds true for other substances. The oil produced by human skin, for example, can be left on a surface in the form of a finger print. At normal pressure, that finger print is stable and essentially benign (unless you've committed a crime!). Yet under a suitable vacuum, just like the water, that oil can be made to boil away.

Now let's suppose that fingerprint was inadvertently left on the interior surface of a vacuum tube. The tube may be highly evacuated and successfully sealed, but as time passes, the oil from that fingerprint will boil away and remain as a vapor within the bulb. The vacuum within the bulb is effectively ruined, because the oil vapor can impede the flow of electrons as well as any atmospheric gas.

Page 135

Some materials, which may be stable at atmospheric pressures, contain constituent substances that are unstable in a vacuum. Certain plastics, for example, contain traces of solvents or plasticizers. In day-to-day life, you'd probably never notice it. In a sealed vacuum, however, those compounds are drawn out of the plastic and into the surrounding vacuum. The emission of gasses from materials exposed to a vacuum is referred to as *outgassing*. Outgassing can ruin a sealed vacuum in short order.

Lest you think metals are immune from this sort of process, consider brass. Although brass has a number of desirable mechanical and electrical properties, its use in vacuum equipment is generally limited to pressures above 0.001 millitorr. This is because brass contains zinc, a metal that's comparatively volatile if warmed to any degree.

Outgassing can occur for reasons other than decomposition, too. One of the alternate causes of outgassing is a process called *adsorption*.

If you take a teaspoon from your silverware drawer and look at it closely, you'd probably think it clean. Yet, despite outward appearances, that spoon is actually encrusted with molecules that don't belong there. The entire surface of the spoon is coated with a molecular film composed of water and various gas molecules. These molecules, which are stuck to the metal, are said to have been *adsorbed* by the spoon. No amount of ordinary cleaning will dislodge or remove them.

Adsorbed molecules can pose a real problem if the host material is part of a sealed vacuum system, like a vacuum tube. A plate or grid with adsorbed molecules will eventually shed some of them. As soon as the adsorbed molecules break free and become mobile again, they join other molecular detritus floating in the bulb. They effectively raise the pressure and degrade the vacuum.

Tube manufacturers went to great lengths to address outgassing. Instead of simply evacuating and sealing the tubes, it was common practice to evacuate them, and then bake them, all the while continuing to run the pumps. Temperatures might be raised to those just below the melting point of the bulb itself! The purpose of this was to drive adsorbed molecules off the interior surface of the glass, and off the other components of the tube. As well, the filament might be powered up and lit. In some cases, induction was used to bring the various metal parts to a dull red glow. All of this was intended to drive off any adsorbed particles so that they couldn't ruin the vacuum later. Only after this torturous treatment would the bulb be sealed once and for all.

Manufacturers employed an additional tactic to combat outgassing. Alkali metals like magnesium, or other substances such as phosphorous, arsenic, or sulfur were vaporized within the bulb. This resulted in a silvery film that collected on the inside of the tube as the vapor condensed on the glass. This film, called a *getter*, had the ability to chemically attack any unwanted molecules that might appear as the result of outgassing. The getter also assisted in sealing gasses within the walls of the tube.

Amateur Vacuum Systems

The ability to draw a vacuum inside of a sealed container affords the amateur scientist or engineering hobbyist endless opportunities for experimentation. Besides thermionic emission, the subject of discussion here, a properly equipped vacuum enthusiast can play with interesting electrical discharges, sputtering, home-built radiometers, and who knows what else.

While it's nice to own brand-new equipment, most of us make do with what we can get our hands on. My own system consists of an old Welch model 1405 rotary oil pump, which was purchased used through an online auction. I paid about 100 dollars, though I've seen similar pumps go for far less. In shipping, it suffered some injury due to poor packaging, but happily, it's so heavy and well designed that most of the damage was cosmetic. A lesser product would probably have been smashed to bits.

On receiving it, I quickly came to the firm conclusion that it was older than I was! Consequently, it required a bit of cleaning and repair to restore it to proper operation. The motor was filthy and caked with gook. It ran well enough, but I was concerned that it might be unable to "breath" properly, which would eventually result in overheating. I pulled the motor apart, cleaned it thoroughly, lubricated the bearings, and then reassembled it.

The base of the pump was spattered with old, grimy pump oil, and there was some evidence that it had leaked from the driveshaft. I removed the packing around the shaft and discovered that the graphite oil seals were worn out and even fractured. I contacted the manufacturer, ordered the parts I needed, and repaired the seal. I did not, however, dismantle the pump itself. This is a fairly involved process, and reassembly requires *new* gaskets and seals. If you're contemplating stripping down a pump of your own, I'd avoid doing it unless you already have the requisite parts on hand, and you're certain there is something to be gained.

I cleaned the exterior of the pump, base, and motor mounts, and then reinstalled the motor with a brand-new drive belt. I refilled the pump with vacuum oil, also purchased from the manufacturer, and the machine ran like a top. There was no longer any indication of leakage, and my total outlay for parts was something on the order of 40 dollars.

A quick search of vacuum-pump dealers on the Internet puts my total cost in perspective. One pump dealer listed a used 1405, without restoration or warranty, for 800 dollars. A new 1405 can set you back as much as 2,000 dollars! When all was said and done, I had secured a high-capacity, high-quality vacuum pump, manufactured by a recognized leader in the industry, for less than 150 dollars.

My workshop is a nice, comfortable, carpeted room. I wanted to move my pump indoors, close to my lab bench, but I had several concerns. First, the 1405 must weigh 100 pounds or more. It's one heavy-duty machine! Second, while the new shaft seals seemed to do the trick, I was apprehensive that a future leak might dump a half-gallon of pump oil on my carpet. Finally, the exhaust from rotary oil pumps is laden with tiny oil droplets. This is not the sort of substance I wanted condensing on the walls and ceiling, let alone the interior of my lungs. I addressed all three concerns in the following manner:

To begin with, I cut a slab of 3/4-inch plywood, sized to match the base of the pump. I attached four heavy-duty casters to the plywood to create a small cart or dolly. With it, the pump can be moved around with little effort.

I decided I needed some kind of tray or basin that could be placed beneath the pump to catch any oil in the event of a leak. During a trip to local hardware store, I found a large, shallow plastic box, sold for storing clothing or shoes beneath one's bed. The box proved to be ideal as an oil catcher, because it was large enough to accept the base of the pump, and deep enough to retain the pump's full complement of oil should the worst ever occur.

The matter of exhaust fumes was addressed with 9 feet of 1-inch-diameter plastic hose. Whenever the pump is to be used, I attach the hose to the pump's exhaust port, and direct the fumes out through an open window. Figure 9.8.

While all of the vacuum tube experiments in this volume were conducted with the aforementioned pump, a spirited email conversation with a fellow experimenter prompted me to think about alternate means of producing suitable vacuums. In essence, he was unhappy that I had chosen to create my vacuum with (yes!) a vacuum pump, because he

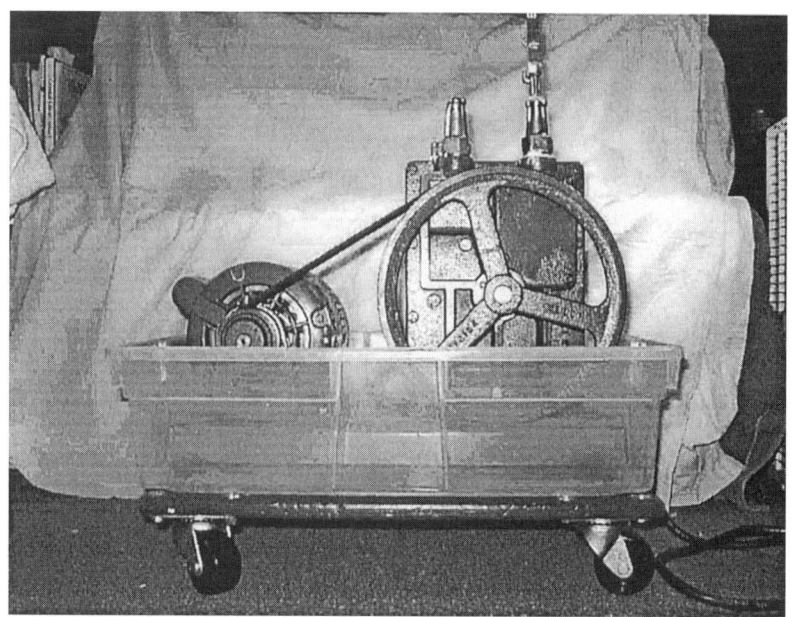

Figure 9.8
The author's vacuum system, a trusty old Welch 1405 double-vane oil pump

felt that this somehow betrayed the "from-scratch" ethic central to this book. In deference to him, let me propose some possible options, with the caveat that I have not explored any of these myself. I'll leave it to you, and him, to determine their relative suitability for your applications.

Stong's book, *The Amateur Scientist*, features an article from experimenter F. B. Lee, in which he describes the construction of a homebuilt atom smasher and its associated vacuum equipment. As the smasher, in essence, amounts to a specialized form of thermionic diode, his vacuum equipment and techniques would probably lend themselves well to thermionic triode experiments.

Lee's vacuum system was entirely "homebrew" in design and construction. The heart of the system was a pair of refrigeration compressors connected in series and used in reverse to create a "rough" vacuum down to 0.1 torr. A mercury-filled glass diffusion pump, also designed from scratch and fabricated by a local glassblower, extended the system's vacuum to a lower limit in the 0.01-millitorr range. All in all, respectable performance, I'd say!

Despite the age of the book (1960), and the probable changes that have occurred in the design and manufacture of modern refrigeration equipment, Lee's design is worth revisiting. While he demonstrates an obvious preference for rotary-type pumps, one of the two pumps in his system was, in fact, a piston pump. He points out, as I did earlier, that piston pumps tend not to work well, unless valving issues are addressed.

Lee's narrative describes the modifications he made to his piston pump, and reported a 50% improvement in its performance. I suspect that refrigeration hardware has changed somewhat since then, so the specifics of his "fix" may no longer be relevant. On the other hand, as long as one understands the root cause of vacuum problems, one can adapt to the hardware at hand.

In any event, I'm certain that suitable compressors could be harvested from modern trash. There's no shortage of dead freezers, refrigerators, or air conditioners. Come to think of it, I suspect that an automotive air-conditioner compressor, driven by an electric motor through a belt, would work too. My guess is that a clever and committed experimenter could scrape together contemporary junk that would perform as well as Lee's equipment.

A few cautions are in order, however. Refrigeration equipment contains gas and fluids under significant pressure. One must be careful not to loosen fittings or couplings before this pressure has been properly relieved. Note that venting refrigerant compounds to the atmosphere is now frowned upon. Refrigerant gases, released into the air, are suspected of facilitating the degeneration of Earth's ozone layer. Reputable mechanics use special evacuation equipment to remove and recover refrigerants from a system before dismantling it.

Another concern I have with refrigeration pumps is the matter of lubrication. In normal use, compressors are kept lubricated by a small quantity of oil that circulates within the system. I wonder if running these pumps "dry" not only diminishes their potential capacity, but may ultimately result in premature failure.

Drawing from the design of the Welch pump, it seems that one might be able to add an oil reservoir, gravity fed, to the output side of the refrigeration pump. Be careful running the pump with no load on the suction side, however. You may blow oil all over the place!

I mentioned earlier that the refrigeration industry uses pumps of their own to evacuate cooling systems and to recover refrigerant gases. These pumps, manufactured by companies like Robinaire and J/B Industries are readily available, and reportedly produce vacuums down to the 20-millitorr level. I've seen these turn up used-on the market from time to time. Nyle Steiner, author of the Web site "Spark, Bang, Buzz and Other Good Stuff," has conducted triode experiments using precisely this sort of equipment.

One of the most ingenious homemade vacuum pumps I've ever seen described appeared in the "Amateur Scientist" column in the August, 1966, issue of *Scientific American*. There, author C. L. Stong describes a pump design by the same Mr. Steiner. The heart of the device is a length of flexible hose or tubing, and a pair of hand-held rollers of the type available at any art or craft store.

One end of the tubing is attached to the chamber one wishes to evacuate. The remainder of the tubing is extended and laid straight upon the surface of a flat board. The diagram accompanying Stong's description shows two arch-top staples, one at each end of the hose, used to secure it to the face of the board.

To run the pump, the operator places a roller on the tubing and pushes down with sufficient force to crush it flat. Maintaining this force, he then moves the roller over the length of the tubing in the direction of its open end. The roller creates a constriction in the bore of the tubing, which moves with the roller, pushing air ahead of it as it travels down the tube.

Before the roller reaches the end of the hose, the operator applies a second roller to the tubing at the original starting point. The second roller is drawn down the pipe like first. The rollers are applied alternately, each beginning its sweep before the other has completed its own. A few drops of oil, used to moisten the interior of the tube, improves the quality of the traveling seal.

Stong likens the motion to "milking a cow." A wave of constrictions traveling down a length of elastic tubing reminds me, instead, of peristalsis, the process by which creatures like humans can swallow.

For simplicity and frugality, a design like this is nearly impossible to improve upon. On the other hand, it occurs to me that the rolling operation could be mechanized.

Figure 9.9 shows the basic layout of a *peristaltic pump*. The pump consists of a flexible tube which is installed to lie in, and follow the contour of, a "U"-shaped housing. Spinning at the enter of the housing is rotor fitted with two spring-loaded rollers. The rollers lie along the circumference of the rotor, and are spaced 180 degrees apart.

The housing and rotor are dimensioned such that the spring-loaded rollers must crush the tubing and form a moving constriction within the bore of the pipe. Using two rollers ensures that at least one constriction is in effect at any given moment, regardless of the rotor's position. The pumping action is smooth and nearly continuous.

In a homebuilt pump, the housing and rotor could probably be fashioned from scraps of wood. A short stack of roller-blade bearings could be used as nice, smooth-running rollers. The rotor could be driven with a simple hand crank or an electric motor geared down to provide slow, high-torque rotation.

Steiner's pump can also be used in conjunction with other pumping equipment. If the free end of the pumping hose is connected to a water aspirator or backwards refrigeration compressor the combined equipment can produce vacuums on the order of 1 millitorr.

There is at least one other method I've come across to generate fairly high vacuums. In this instance, the pump contains no moving parts whatsoever. It functions through the process called *adsorption.*

An adsorption pump consists of a container, metal or glass, which is filled with a material called *molecular sieve.* Molecular sieve material is composed of tiny pellets that are so riddled with pores and crevices that a single pellet may have the equivalent surface area of *thousands* of square meters. Under the right conditions, gas molecules like to cling to these pores.

To prepare the pump for use, the container, filled with the sieve material, is heated to 350 degrees C. This provides sufficient energy to drive off water vapor and any trapped molecules. Once the sieve material has been purged, the container can be connected to the system one wishes to evacuate.

Operation of the pump is curiously simple. Pumping commences when the molecular sieve container is immersed in a beverage cooler filled with liquid nitrogen. Reportedly, the pump is capable of creating vacuums down to 10 millitorr. The sieve material, by the way, is re-useable. It can be reheated to drive off captured gases, and then used again.

A more specific discussion of the technique, including sources for materials and appropriate safety concerns, appears in the article "Working in a Vacuum" by Dr. Shawn Carlson. It was featured in the "Amateur Scientist" column in October of 1996.

I should mention that all of the excellent "Amateur Scientist" columns, which have appeared in *Scientific American* from 1928 to 1999, are now available in a comprehensive CD-ROM. See my references for specific information. I have found the material extremely interesting and useful. If nothing else, the purchase of that CD-ROM has saved me untold trips to the library.

You may note that I've not elaborated on the subject of diffusion pumps. I, for one, found sufficient opportunity for experimentation without one, though I understand a more sophisticated tinkerer might insist on the enhanced capabilities it can provide. Yes, diffusion pumps can be built from scratch, and have been. On the other hand, it's much easier to build a poor diffusion pump than a good one. I can see little point in expending that sort of effort when high-quality used pumps regularly appear on Internet auction sites. To each his own, I guess.

I would, however, discourage anyone from using mercury as a pumping medium. In this day and age, local officials absolutely freak out over the tiny beads that escape from a broken thermometer. (I seem to recall a news story where a school was actually *closed* because somebody had dropped a thermometer.) True, mercury in its liquid form is relatively benign if not ingested. However, mercury vapor is *extremely* toxic as it directly attacks the lungs and nervous system.

A mercury diffusion pump requires a fair amount of the metal. As you now know, diffusion pumps contain heaters whose purpose it is to vaporize the pumping medium. All said, a malfunction or leak in a system of this type can pose considerable health risks. If you *insist* on using a diffusion pump, I can see no reason to use mercury when numerous oil-based pumping compounds are readily available.

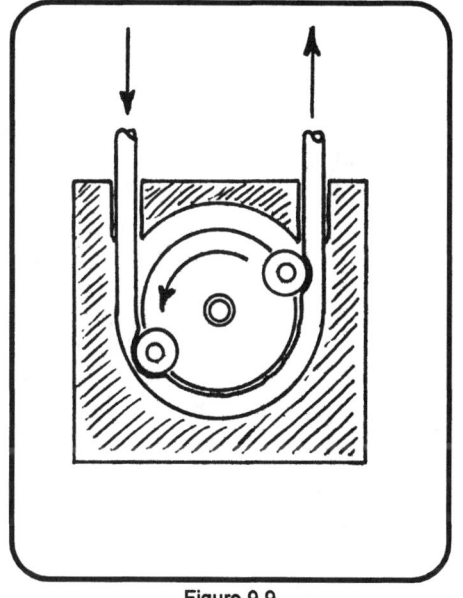

Figure 9.9
A peristaltic pump

Amateur Vacuum Measurements

In the worlds of professional science and engineering, nothing is more important than the tools one uses to measure with. A professional-grade measurement instrument must have three very important characteristics.

First, it must be *accurate*. In other words, the tool must report meaningful values when measuring a given quantity. Accuracy is usually ensured by a system of calibration, where the tool is compared to a standard to ensure its readings are appropriate for the quantities being measured.

Next, a measurement instrument must be *precise*. If a precise instrument is used to make repeated measurements of a single, fixed quantity, it will always report the same value. Precise measurements are repeatable.

When professional instruments are calibrated, special records are created to document which standards were used in the calibration process. Frequently, measurement standards are themselves calibrated against other, more authoritative standards. Similar calibration records are maintained for them as well.

All of this record keeping establishes a formal paper trail or lineage that can link a measurement instrument to international standards. Thus, if the legitimacy of an instrument's measurement is ever called into question, one can defend it by virtue of the specific standards used to calibrate it.

The ability of an instrument's measurements to be validated on the basis of calibration records is called *traceability*, and it's the third characteristic of a professional measuring instrument.

Mind you, these attributes don't come cheap. Thankfully, most amateur work is sufficiently informal that these stringent requirements can be relaxed. This means that suitable measurement equipment can be found, salvaged, or built. This holds true for the vacuum-related projects depicted in this book. Accurate pressure measurements are desirable, but not crucial to the thermionic projects to follow.

The simplest homebuilt pressure-measuring device is the classic manometer, the design of which was covered in detail some pages back. Vacuum levels in Lee's atom smasher, for example, were measured with manometers and a variant of the manometer called a *McLeod gauge*. McLeod gauges are capable of measuring high vacuums with good accuracy. However, the extensive use of mercury in all of these instruments makes them, in my opinion, a poor choice for casual experimentation.

My measurement system is based on a Varian Model 801 millitorr vacuum gauge. The Model 801 is composed of a panel-mount meter, a thermocouple gauge tube, and a cable that links the two. The gauge tube is the actual sensing element, detecting gas pressure thermally. The 801 has an advertised measurement range of 2 torr down to 1 millitorr. See figure 9.10.

My gauge was salvaged from junk, costing virtually nothing. If you're not quite that lucky, such instrumentation may be purchased used. Equipment of this type also appears with some regularity on Internet auction sites. You can search for used laboratory gear, but don't overlook the equipment used by the folks who service refrigeration equipment.

A comparable instrument, believe it or not, can be built from scratch. In the "Amateur Scientist" column dated November, 1996, Dr. Carlson describes a homebuilt Pirani gauge created by Bruce R. Kendall, at Pennsylvania State University. The key to this ingenious design is a glow plug, commonly used on the tiny engines of radio-controlled airplanes. The plug is heated by an electric current from a simple circuit that regulates the flow, limiting it to a fixed value of 1.4 amperes. (It appears this value was chosen to be compatible with the specific glow plug used, an O. S. Engine Model A5. I suspect the precise value is not as important as the fact that it's maintained at a constant level.) Additional circuitry, in the form of an instrumentation amplifier, produces a voltage representing the measured pressure.

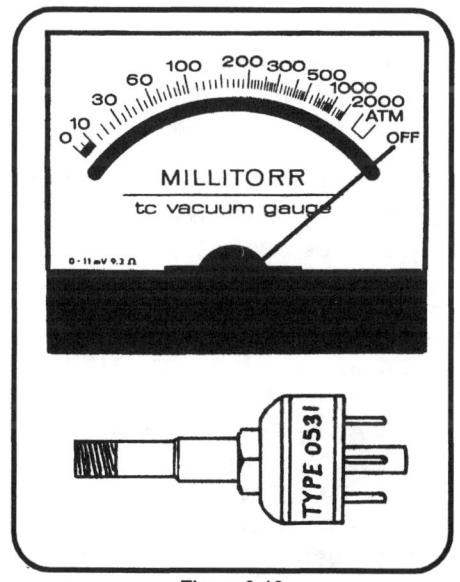

Figure 9.10
The author's thermocouple vacuum gauge

References

Appleton, E. V., *Thermionic Vacuum Tubes*, copyright 1931, E. P. Dutton and Company, New York, pp. 5-14, (construction of vacuum tubes)

Carlson, Shawn (editor), *Scientific American's "The Amateur Scientist " - The Complete Collection on CD-ROM*, copyright 2000, The Tinker's Guild, Menlo Park, CA, (multiple references to "Amateur Scientist" articles)

Martin, L.H., and Hill, R.D., *A Manual of Vacuum Practice*, copyright 1947, Melbourne University Press, Australia, reprinted by Lindsay Publications, copyright 1997, ISBN 1-55918-188-5, pp. 15-38, (pressure-measurement techniques), pp. 39-63, (pumps), pp. 64-69 (materials used in vacuum systems), pp. 83-90, (metal to glass seals), pp. 101-105, (gasses in metals and glass)

Moore, John H., Davis, Christopher C., and Coplan, Michael A., *Building Scientific Apparatus*, copyright 1989, Addison-Wesley Publishing Company, Reading, Massachusetts, ISBN 0-201-13189-7, pp. 80-83, (pressure measurement techniques), pp. 83-90, (vacuum pumps), p. 91, (volatility of zinc in vacuums greater than 10^{-6} torr)

Moyer, James A., and Wostrel, John F., *Radio Receiving Tubes*, copyright 1931, The Maple Press Company, York, PA, pp. 1-9, (history of the vacuum tube), pp. 9-23, (construction of radio tubes)

Radio Corporation of America, *RCA Receiving Tube Manual*, copyright 1959, Radio Corporation of America, Harrison, N.J., p. 325, (materials used in construction of tubes, and vacuum references)

Stong, C. L., *The Amateur Scientist*, copyright 1960, Simon and Schuster, Inc., New York, N.Y., pp. 352-355, (vacuum pumps built from refrigerator compressors)

Strong, John, *Procedures in Experimental Physics*, copyright 1938, Prentice-Hall, Inc., New York, reprinted by Lindsay Publications, copyright 1986, ISBN 0-917914-56-2, pp. 111-122, (designs for diffusion pumps), pp. 123-124, (virtual leaks), pp. 130-131, (electrodes), pp.137-150 (vacuum gauges)

White, Harvey E., *Modern College Physics*, copyright 1956, D. Van Nostrand Company, Inc., New Jersey, pp. 580-582, (electrical discharges in the Crookes tube)

http://www.omega.com/literature/transactions/volume3/high2.html
"High Pressure and Vacuum - Vacuum Measurement," copyright 2000, Omega Engineering, Inc.

http://www.industrysearch.com.au/features/vacuum.asp
"Vacuum Science: A User's Guide to Vacuum Technology," copyright 2001, IndustrySearch.com.au, (discussion of the diffusion pump, and ranges and applications for vacuums)

http://www.varianinc.com/vacuum/products/gauging.html
"Varian Vacuum Technologies - Gauges and Instruments," copyright 2000, Varian, Inc.

http://home.earthlink.net/~lenyr/
Steiner, Nyle, "Spark, Bang, Buzz and Other Good Stuff," copyright ?, last updated March, 2001, (vacuum tube experiments and vacuum equipment)

http://www.tiac.net/users/shansen/belljar/pump.htm
Hansen, Steve (editor) "Refrigeration Service Vacuum Pumps," copyright 1996, originally published in Volume 4, Number 1 of the *Bell Jar*, ISSN 1071-4219, (use of refrigeration pumps for vacuum experimentation)

Chapter 10
Experiments With
Glow Tubes and Diodes

With my simple vacuum system up and running, I decided to conduct a few simple experiments to see if my equipment would produce a suitable vacuum to allow for measurable thermionic emission. It was also my goal to test some ideas regarding materials and techniques for the vacuum tubes I expected to create later on.

The Spice Jar Glow Tube

Dry air, under normal atmospheric pressure, is not a particularly good conductor of electricity. If, however, a voltage is applied that's high enough, electrons can be violently ripped from the atoms comprising the gas, and an electric current can force its way through it. It may take as much as 30,000 volts to cause a spark to jump less than a half inch. This should give you some clue as to the immense voltages at work in the monstrous electrical discharge called *lightning*.

The glow generated when electricity passes through a gas is due to the excitation of atoms. (If necessary, review the introduction to atomic structure a few chapters back.) At reduced pressures, less voltage is require to elicit the breakdown of the gas, and if the pressure is low enough, the visible discharge takes on some interesting properties.

A 6-inch glass tube connected to a vacuum pump and a high-voltage supply on the order of 10,000 - 15,000 volts begins to display blue streamers at a pressure of 10 torr. At 5 torr, the streamers vanish

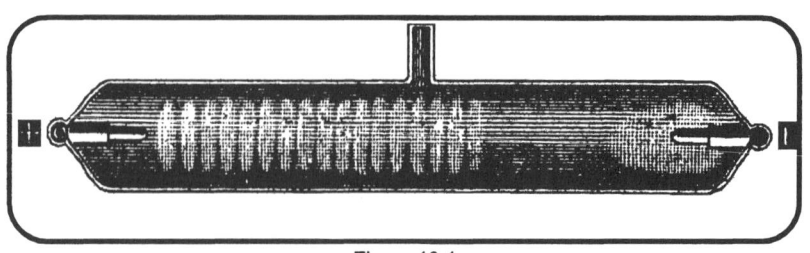

Figure 10.1
Patterns appearing in a glow tube
(Adapted from *Electricity and Magnetism*, 1905)

and are replaced with a uniform pink glow. At 2 torr, a dark space begins to appear at the negative end of the glowing column. This is referred to as the *Faraday dark space*. Also, at this stage, the column changes hue, growing more blue near the negative end, while remaining pink towards the positive.

Still harder vacuums produce dark striations or stripes through the column, and a second dark region forms around the negative electrode itself. This latter dark region is called the *Crookes dark space*. See figure 10.1.

As pumping continues, the Crookes space grows until, around 10 millitorr, the entire tube darkens. At this stage, the glass itself may glow a faint green color. This fluorescence is due to the stimulation of atoms in the glass, which are bombarded and excited by electrons racing through the tube. If the bulb blacks out entirely, this is an indication of a vacuum on the order of 1 millitorr or less.

The glow tube depicted in figure 10.1 is typical of the equipment that might be found in a demonstration or teaching lab. It's fashioned from blown glass with wire electrodes sealed at each end. While these tubes are beautiful, they're expensive. They require glassblowing equipment and materials that makes home workshop replication a non-trivial task.

As part of the process of evaluating my pump, I fashioned my own glow tube from a small spice bottle. See figure 10.2. The bottle, roughly 3-1/2 inches tall and 1-1/2 inches in diameter, used to contain garlic power, cinnamon, or similar spices. It's typical of the kind sold in supermarkets everywhere.

The neck of the bottle was fitted with a rubber stopper, which was bored to accept a 4-inch length of 1/4-inch copper tubing. The copper tubing, terminated with a compression fitting on its free end, provides not only an airtight connection to my pump, but also acts as one of the tube's electrodes.

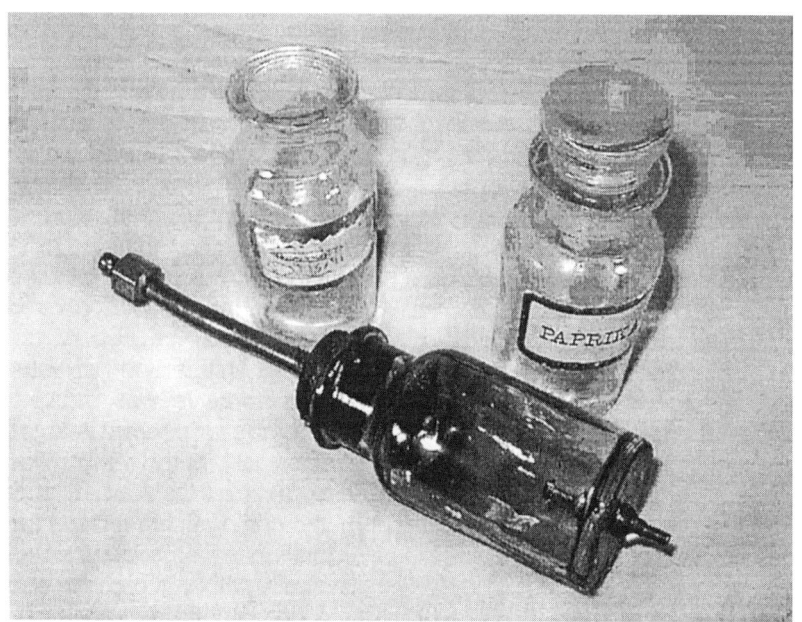

Figure 10.2
The Spice Jar Glow Tube

In boring the stopper, I should note that a common twist drill bit is really not a good tool to use. Rubber is compliant, so it's prone to tear and shred as opposed to cutting cleanly. A hole drilled with a regular bit is liable to have rough and ragged walls which, when fitted with the copper tube, tends to leak.

I prefer to use a short length of brass tubing, sharpened at the cutting end, as my drilling tool. The tubing works best if chucked in a drill press and rotated at a relatively slow speed. When used this way, the brass tube bores a clean hole with fairly smooth walls.

If you look closely at figure 10.3, you'll note that the second electrode is actually a 6-32 brass machine screw that pierces the bottom of the bottle. The screw I used was 1-1/2 inches in length. It was fitted with a brass nut, which was tightened to within 1/2 inch of the screw's head. There, it was soldered in place. Then, the side of the nut facing away from the head, along with the adjacent screw threads, were coated with a glob of quick-setting epoxy. Using forceps, I lowered the screw through the neck of the bottle, and inserted it through a 1/8-inch hole previously drilled in the floor of the bottle. I spun on the second nut, threaded from the outside of the bottle, and the electrode was complete. The nuts provide structural integrity, while the epoxy creates an acceptable seal. Be careful not to over tighten the outside nut, or you may risk cracking the bottle.

The fact that the second electrode requires a hole in the floor of the bottle begs the question, "how do you drill a hole in glass?" A classic, tried-and-true method appears in numerous laboratory texts. In the old days, a piece of brass or copper tubing, slightly smaller in diameter than the desired hole, was chucked into a drill press. The end of the tube was nicked with a sharp blade, and embedded with carborundum or silicon carbide grit. Drilling was really a grinding operation. The improvised bit was rotated against the glass, while a slurry of water and abrasive was dribbled onto the work area. A small wax or putty dam was sometimes fashioned to pool the cutting fluids around the drill site. I experimented briefly with this process and found it both frustrating and time consuming.

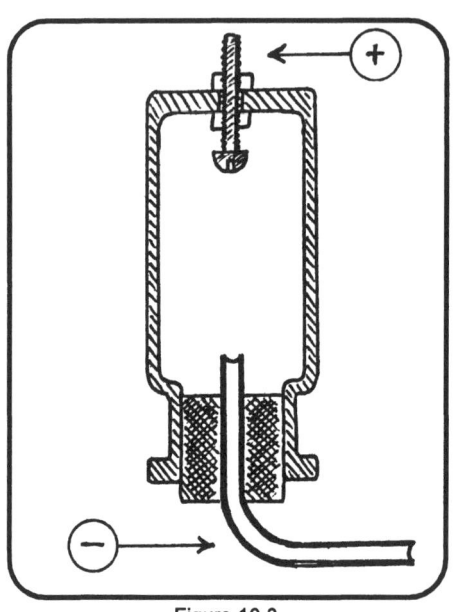

Figure 10.3
Cutaway of the Spice Jar Glow Tube, showing the relationship between various components

Recently, tool manufacturers have introduced special purpose drill bits intended for working glass, ceramics, and other, exceptionally hard materials. These bits have the appearance of a bullet-shaped paddle, welded to a steel shaft or shank. They're fairly expensive at five or six dollars apiece, but they work very well and are much faster than traditional drilling methods. Happily, these bits can now be found in many hardware stores and home-improvement centers.

Let me share with you three tips for their successful use: First, the bits can probably be used in a hand drill, but I strongly recommend a drill press. Glass is a brittle and unforgiving material. The drill press allows for more precise control of the bit. Second, use the slowest speed possible. Higher speeds will simply heat and burn up the bit. Finally, I like to drip a light grade of machine oil on the bit as I drill. I've

Figure 10.4
An assortment of induction coils for producing high voltage

found that the bit seems to cut better, it remains cool, and the oil controls the glass dust produced by the drilling operation. By the way, make sure you're wearing safety goggles and a dust mask.

The stopper, copper pipe, and bottle will assemble just fine without any additional sealants. I found, however, that a dab of contact cement here and there can greatly improved the integrity of the seal. If you moisten the end of the copper tube with cement before inserting it into the stopper, you'll find it acts as a lubricant. The tube will slide into the stopper with ease, then the joint becomes virtually permanent. The same can be done with the stopper itself, before it's inserted into the bottle. Beware, you may never get the stopper out again.

A glow tube requires a power supply capable of delivering several thousand volts. How can you generate that kind of voltage?

One way to power glow tubes is with an *induction coil*. An induction coil is a type of transformer that raises battery voltages to many thousands of volts. It's similar to the ignition coils found in automobiles, except that it contains a built-in set of vibrating contacts, called a *trembler*, to interrupt the primary current. (Review the chapter on transformers to understand why this is necessary.)

Figure 10.4 shows a small assortment of induction coils. The unit to the far right is a vintage coil used as part of the ignition system for a Ford Model T. I picked it up at a flea market a few years back for less than 10 dollars. Similar units are still manufactured for antique-auto enthusiasts, and are available through mail order. The middle coil is also a genuine antique, probably used for an early engine ignition application. The largest is representative of the laboratory-grade coils that are still being manufactured and are readily available. The cheapest coils I've seen start at 50 dollars, while larger and more ornate units can exceed 1,500 dollars. Any of them will work just fine.

A cheaper power source is a common automobile ignition coil. I'm speaking specifically of the can-shaped variety, with the two screw terminals and the raised plastic nipple on top. You may find one of these at a garage sale for a couple of quarters, or buy one new for less than 20 dollars. Note that modern coils do not have a trembler. Alone, they cannot produce a continuous stream of sparks like the induction coils can.

Modern ignition coils can be induced to produce high-voltage pulses with a simple push button. A single, high-voltage spark will result each time the button is pressed and released. An external trembler, either mechanical or electronic, must be used to develop a more continuous stream of sparks.

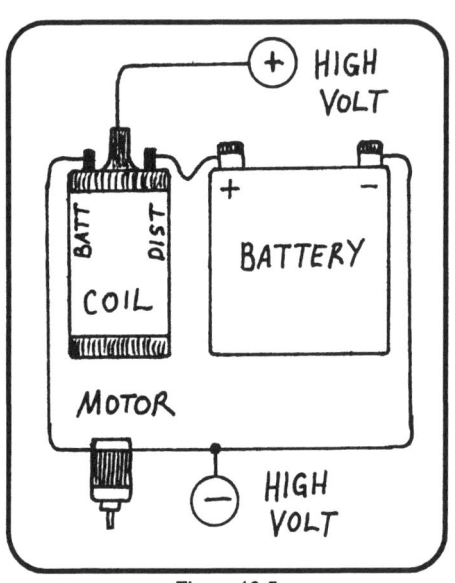

Figure 10.5
Producing high voltage with an old automotive ignition coil

A quick and dirty alternative to a trembler is a low-voltage DC buzzer or low-voltage DC motor wired in series with the primary of the coil. The internal switch contacts (in the case of the buzzer) or the commutator (in the case of the motor) act to make and break the circuit, which aids the coil in doing its thing. See figure 10.5.

A lit glow tube is very beautiful and fun to play with. Because the glowing matter inside consists of charged particles, they will react to the influence of an external magnet. The luminescent haze inside the bottle can be distorted, constricted, and squashed, depending on the nature of the applied magnetic field. If you've never witnessed the Aurora Borealis, this is an opportunity to see it up close. The Northern Lights, like the clouds in the glow tube, are composed of excited gas atoms under the influence of Earth's magnetic field.

A couple of warnings are in order here. Spice bottles were not intended to be used as glow tubes. When you evacuate them, they're subjected to a significant crushing force from the atmosphere. Always wear eye projection when working with evacuated glass, just in case a bottle or container should implode. A two-liter soda bottle, cut down with scissors, makes a nice, clear, plastic shield to cover the tube while it's in operation. Don't use it in place of goggles, use it in addition to them.

The second warning deals with the power supply. Induction and ignition coils produce voltages that, while generally not lethal, can certainly burn you. Also, an electric spark jumping into the hand of a careless experimenter may produce involuntary muscle contractions leading to a fall or other injury. Play it safe. Make sure your electrical connections are well insulated. Make sure the power is off before making any changes to the wiring.

So, what's does this equipment have to do with building vacuum tubes? Well, if you've constructed a vacuum system without a means to measure system pressure, a glow tube can be used to give you a rough estimate.

The pressure gauge I salvaged for use with my system was not calibrated. I had no idea if its readings were meaningful, or corrupted and useless. By comparing the readings on the gauge to the behavior of my glow tube, I was able to gain confidence that its measurements were in the right ballpark. The best vacuum I achieved nearly blackened the interior of the bulb, while the glass glowed green. I figure the actual vacuum was probably slightly better than the 15 millitorrs displayed by the gauge. Certainly, if your system can remove enough air to completely blacken a glow tube, then you definitely have the capacity to fabricate experimental triodes.

The Canning Jar Diode

The next logical step, I decided, was to attempt to build a simple thermionic diode. I reasoned that if I couldn't produce measurable thermionic emission in a diode, there was little point in attempting to build a triode.

The diode was not a project that I cared to invest a great deal of effort in, so I avoided labor-intensive designs. I was looking for something, fashioned from common materials, that I could knock together quickly. Fully expecting burned filaments, I also wanted a design that would allow the filaments to be replaced. I call the end result the "Canning Jar Diode." It can be seen in figure 10.6.

The body of the diode is a common canning jar, like those available at any supermarket. A 3/4-inch hole was drilled in the lid to admit a suitable rubber stopper. The stopper itself was bored and fitted with a short length of 1/4-inch copper tubing. The tubing was terminated with a compression fitting, to allow a gas-tight connection to my vacuum equipment. Contact cement was used to ensure a seal, though RTV silicone would have worked as well or better.

I suppose a simpler approach would be to drill a hole in the lid just large enough to admit the copper tube. The copper tube could then be soldered directly to the lid. If you elect to go this route, make sure to sand the enamel off the lid near the joint, or the solder won't stick. You'll also have to be judicious in the application of heat. Too much, and you'll end up burning the gasket on the lid.

A 1/8-inch hole was bored through the center of the floor of the jar using the drill and technique described in the glow tube text. Instead of using a 6-32 bolt for a terminal, I used a 1-1/2-inch length of 6-32 threaded brass rod. The rod was installed in the hole, secured, and sealed in the same manner as described for the construction of the glow tube. Roughly an inch of threaded rod was left protruding into the interior of the jar.

The plate assembly in my diode was composed of a brass disk, about 2 inches in diameter. I soldered a 6-32 threaded sleeve or standoff to the center of the plate and then screwed the plate assembly onto the threaded rod protruding from the bottom of the jar.

I could have installed the plate through more permanent means, but the threads are nice because they allow for adjustability. Rotating the plate screws it up and down the threaded rod, thereby changing the distance from the plate to the filament.

The filament terminals pass through the sides of the glass. The usual techniques were used to drill the holes and secure and seal 6-32 threaded rod through the walls of the jar. A pair of brass jam nuts were threaded onto each filament terminal inside the jar. These nuts were intended to facilitate installing and connecting a filament. See figure 10.7.

I already knew that many early vacuum tubes used tungsten as a filament material, though it wasn't immediately evident to me where

Figure 10.6
The Canning Jar Diode, an experimental thermionic diode

I might find wire of that type. I did have a good bit of fine-gauge resistance wire (containing nickel) on hand, and thought it might prove a suitable substitute. Resistance wire is intended for use in, among other things, heaters. I expected it to be tolerant of high heat, even more so in a vacuum where the wire can't burn. It seemed an ideal filament material. I cut a short length, and connected it to the filament terminals using the 6-32 nuts to make good, solid connections.

The diode was hooked up according to figure 10.8. With this arrangement, I was able to measure and vary the filament current, while monitoring its effect on the plate current. The plate supply was a 90-volt battery composed of a stack of 10 9-volt transistor batteries wired together in series. The jar was pumped down to about 35 millitorr, the filament was lit, and testing commenced.

To be blunt, I was quite disappointed. Starting with a dull red glow, I slowly increased the voltage applied to the filament. It grew brighter and brighter, but there was no measurable plate current. In fact, the only evidence of thermionic emission occurred just before the filament disintegrated. I raised the filament current as high as I could; I saw a few microamps of plate current, and then the jar went dark. I tried several other materials, including steel wire, iron wire, and copper,

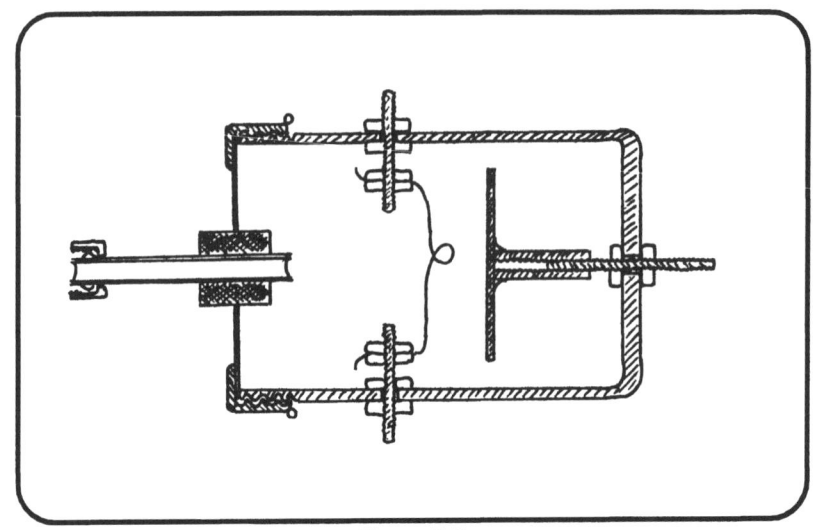

Figure 10.7
A cutaway view of the Canning Jar Diode showing the relationship between
components

without success. None of these were particularly good emitters, either
due to their respective work functions, or to the fact that they couldn't be
raised to a suitable operating temperature without self-destructing.

In the context of work functions of these metals, some of the
failures make perfect sense. Had I considered this at the time, I might
have saved myself some time and effort. On the other hand, I take
pleasure in seeing a theoretical detail mesh well with real-world
observations, so I'm glad I took the trouble to play with the alternative
materials. In the end, it was clear that I needed a tungsten filament. The
question was, where would I find one?

An obvious source of tungsten filaments are common light
bulbs. I actually purchased several different types for the expressed
purpose of breaking them and removing their filaments. Among my
favorites are the incandescent bulbs intended for use in aquarium
hoods. (Please be clear on this...I'm *not* talking about the fluorescent
variety!) Incandescent aquarium bulbs are low-wattage lamps
containing long, thin, straight filaments that are easy to harvest.

Rather than simply smashing the glass, which would risk
damaging the filament, I scored the circumference of the bulb near the
base. A glass cutter will work, though I used the broken point of a small
triangular file. A gentle tap with the point of a screwdriver caused the

glass to cleave, and the glass bulb was lifted away intact. I would recommend wearing leather gloves and safety glasses during this operation, and performing the work inside of an old cardboard box. The box will retain all the shards of broken glass should something go wrong.

Tungsten filaments, it turns out, don't lend themselves well to handling, particularly with one's fingers. They also proved nearly impossible to connect directly to the filament terminals inside the diode.

To overcome this, I cut two short lengths of bare, copper wire, about 24 gauge, and attached them to the filament terminals inside the jar. The very tip of each wire was bent into a sharp "V"-shape, or hook. Next, using tweezers, I manipulated the ends of the filament into the mouth of each hook. Needle-nosed pliers were used to the crush the hook closed, thereby crimping the copper lead onto the tungsten. Figure 10.9 offers some detail as to how this is done.

The tables near the end of this chapter depict a series of measurements made to document the relationship between filament current (which affects its temperature) and plate current. Note that at its best, the diode with the

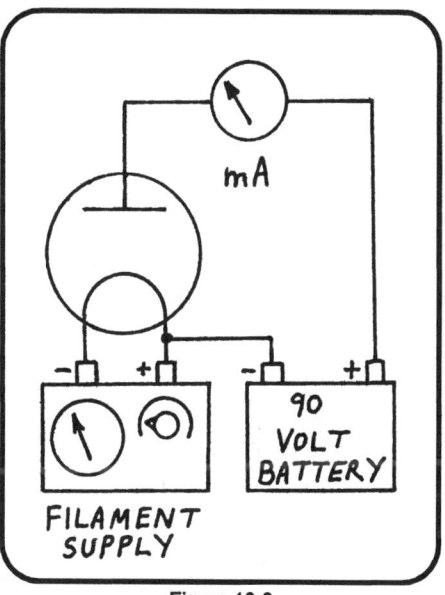

Figure 10.8
Measuring the behavior of an experimental thermionic diode

aquarium tube filament passed a respectable 13 milliamperes of current, with an effective forward resistance of roughly 7,000 ohms. If the plate battery was reversed, plate current dropped to zero, as expected. Yes, this diode actually worked!

I tried an assortment of filaments extracted from a wide variety of incandescent lamps. One of the more successful tests involved a filament taken from a cartridge-type automobile courtesy-lamp bulb. The donor bulb was carefully crushed with slip-joint pliers, and the filament was extracted. The second data table shows that the diode with the automotive filament was able to carry larger currents, though I'm not quite certain why. Just prior to burnout, the filament was driven to a brilliant white heat by 0.65 amperes. The resulting plate current was about 118 milliamperes, and the effective resistance was a mere 739 ohms! Perhaps there was something about the filament's composition that made it an inherently better thermionic emitter.

The power dissipated by a diode during this type of test can be calculated by multiplying the plate voltage times the plate current. During the experiment with the automotive filament, the canning jar diode dissipated in excess of 10 watts!

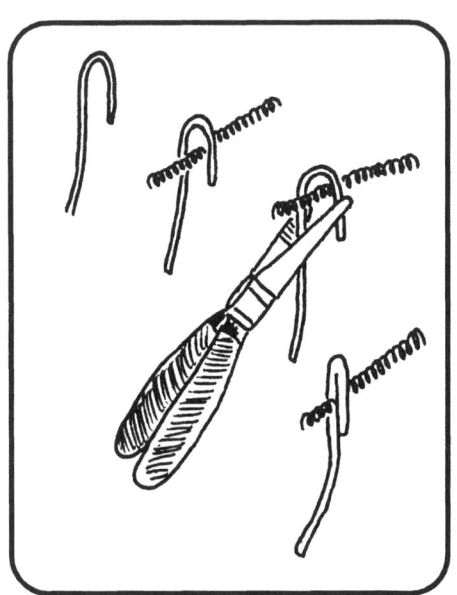

Figure 10.9
Joining copper lead wires to a tungsten filament

If you look closely at the data tables, you'll notice that the relationship between the filament and plate currents is not linear. Useable emission begins only at the highest filament currents, where the tungsten is heated to yellow-white incandescence. As the filament currents are raised, and the filament is driven closer and closer to the point of destruction, electron emission skyrockets until the tungsten disintegrates and the circuit is broken.

The downside of this is that the hotter a filament is run, the shorter its useful life. This rule applies to all incandescent light bulbs, and it's particularly true for sections of tungsten filament that have been subjected to the mechanical shock of removal from their hosts, handling and manipulation, and reinstallation into your equipment.

The users of homemade vacuum tubes based on this design must draw a line of compromise between useable plate current and useable filament life. Be warned, my luck generally ran out after 10 - 15 minutes of operation.

Canning Jar
Vacuum Diode

Pressure:	35 millitorr	
Filament Material:	Tungsten, removed from	120-volt, 25-watt
	tubular Edison-based lamp	
Plate Supply:	Battery, 90 volts nominal	

Filament Current (A)	Plate Current (mA)	Plate Voltage (V)	Forward Resistance (Ohms)
0.04	0.0001	94.4	944,000,000
0.05	0.0008	94.4	118,000,000
0.06	0.008	94.4	11,238,095
0.07	0.025	94.4	3,806,452
0.08	0.130	94.2	724,615
0.09	0.170	94.2	554,118
0.10	0.570	93.8	164,561
0.11	3.400	92.5	27,206
0.12	5.750	91.8	15,965
0.13	11.750	91.5	7,787
0.14	13.260	89.9	6,780
Burnout			

Canning Jar
Vacuum Diode

Pressure:	35 millitorr
Filament Material:	Tungsten, removed from 12-volt cartridge-type automotive dome lamp
Plate Supply:	Battery, 90 volts nominal

Filament Current (A)	Plate Current (mA)	Plate Voltage (V)	Forward Resistance (Ohms)
0.23	0.0002	96.9	484,500,000
0.30	0.008	96.7	12,240,506
0.35	0.088	96.2	1,093,182
0.38	0.337	95.7	283,976
0.40	0.543	95.4	175,691
0.41	1.050	95.1	90,571
0.42	1.210	95.1	78,595
0.43	1.670	94.9	56,826
0.44	2.150	94.7	44,047
0.45	2.860	94.4	33,007
0.46	3.720	94.2	25,323
0.47	4.080	94.1	23,064
0.48	4.540	94.0	20,705
0.49	5.010	93.9	18,743
0.52	3.320	94.5	28,464
0.53	4.060	94.2	23,202
0.54	4.550	94.0	20,659

0.55	6.730	93.4	13,878
0.56	7.300	93.3	12,781
0.57	11.570	92.2	7,969
0.58	12.960	91.8	7,083
0.60	36.900	87.4	2,369
0.65	118.000	87.2	739
Burnout			

References

Martin, L.H., and Hill, R.D., *A Manual of Vacuum Practice*, copyright 1947, Melbourne University Press, Australia, reprinted by Lindsay Publications, copyright 1997, ISBN 1-55918-188-5, pp. 15-16, (attributes of gas discharges, gas discharges as a measuring technique)

Moore, John H., Davis, Christopher C., and Coplan, Michael A., *Building Scientific Apparatus*, copyright 1989, Addison-Wesley Publishing Company, Reading, Massachusetts, ISBN 0-201-13189-7, pp. 72-73, (drilling holes in glass)

Strong, John, *Procedures in Experimental Physics*, copyright 1938, Prentice-Hall, Inc., New York, reprinted by Lindsay Publications, copyright 1986, ISBN 0-917914-56-2, pp. 36-37, (drilling holes in glass)

Thompson, Silvanus P., *Electricity and Magnetism*, copyright 1905? Thompson and Thomas, Chicago, p. 432, (glow discharges in glass tubes)

White, Harvey E., *Modern College Physics*, copyright 1956, D. Van Nostrand Company, Inc., New Jersey, pp. 580-582, (electrical discharges in the Crookes tube)

Chapter 11
The Bell Jar Audion

I n the previous chapters we discussed, on an introductory level, the nature of the atom, the process of thermionic emission, and the origin of the thermionic triode. This was followed by a necessary discussion of vacuums, and the equipment both to produce and measure them. Here, at last, is where these details merge to produce a functional thermionic amplification device.

Lee DeForest introduced the first thermionic triode in 1907, calling it the "Audion." DeForest's Audion, at first glance, strongly resembled a globe-like lightbulb, complete with a screw-in base. It differed from a lightbulb, of course, with the addition of a zigzag grid wire, and the metal plate, which entered the bulb from the top. See figure 11.1. This chapter describes the construction of a homebuilt triode I call the Bell Jar Audion, inspired by and named in deference to DeForest's device. See figure 11.2.

The filament, grid, and plate in DeForest's tube were arranged in a planar fashion; that is to say, they formed a sandwich, with the grid separating the filament and plate. My Bell Jar Audion features a similar electrode geometry. Planar construction, in my opinion, simplifies many of the mechanical details associated with the construction of triodes.

DeForest's tubes were blown from glass, evacuated, and sealed. The vacuum inside his tubes was static, intended to be more or less permanent. For me, this made replication problematic. I had no equipment (torches, annealing ovens, etc.) with which to properly work glass. While I could easily envision a hundred alternative methods and materials from which to build my own tube, virtually all of them carried certain risks of outgassing. This meant that I could go through the motions of evacuating and sealing my triode, but in short order, that vacuum would be ruined. In the end, I opted to press ahead with the alternative materials, and simply dispense with the idea of ever sealing the tube. The vacuum in my triode would be dynamic, generated by my vacuum pump as needed.

All in all, I don't consider a dynamic vacuum a particularly grave concession. The Bell Jar Audion is an experimental device and makes no pretense of being a replacement for any commercially produced vacuum tube.

Let's now take a closer look at some of the actual construction details associated with the Bell Jar Audion (BJA for short).

The Base

The base of the BJA was composed of a disk of aluminum, 4 inches in diameter and 1/8 inch thick. There's nothing significant about either of these dimensions beyond the fact that my junk box already contained this disk as a castoff from another project. However, because of the drilling and threading operations that follow, I wouldn't use aluminum stock thinner than 1/8 inch.

I located the center of the disk and then, using a compass, I scribed a circle about 1-1/2 inches in diameter. Using the same setting on the compass, I walked it along the circumference of that circle to subdivide it into six equally spaced points. I used a 1/8-inch bit to drill a pilot hole at each point.

Next, the six holes were enlarged with an 11/32-inch bit, and then threaded with a 1/8" NPT tap. NPT taps are designed to cut threads for pipe fittings, and are available at any decent hardware store.

Note that pipe threads differ from ordinary machine-screw threads in an important detail that bears some explanation. A hole that's been threaded for an ordinary bolt or machine screw is uniform in diameter from one end to the other. This means you can insert a short length of threaded rod and drive it all the way through. NPT threads, on the other hand, are not uniform. In fact, their diameter is slightly tapered. The farther you travel into an NPT tapped hole, the smaller the diameter gets.

When a piece of threaded pipe is screwed into a fitting and tightened, the tapered hole causes a wedging or squeezing action that helps to effect a water-tight or gas-tight seal.

This is a crucial point to keep in mind as the BJA's base is threaded. The tap should be driven into the holes just far enough to cut a complete set of threads. Don't drive the tap all the way through! To do so will ruin the taper and make it virtually impossible to properly seal any fittings screwed into those holes.

One of the six threaded holes was selected as an evacuation port. I installed a brass, 1/8-inch NPT 90-degree elbow and then fitted it with a 2-inch brass nipple. The end of the nipple was terminated with a compression fitting that would allow the BJA to be linked to my vacuum pump with 1/4-inch copper tubing. At each joint, the pipe threads were liberally coated with teflon pipe sealant (paste) before

Figure 11.1
An example of DeForest's Audion
(Courtesy of John Jenkins
www.sparkmuseum.com)

being tightened.

Finally, I set the compass to scribe another circle, this one about 1/4-inch (radius) smaller than the diameter of the disk itself. Somewhere along that circle I drilled two 1/8-inch holes, spaced at

Figure 11.2
The Bell Jar Audion

opposite ends of the disk. These would be used later as a means by which to mount the triode to its pedestal or stand. See figure 11.3.

Feed-Throughs

A directly heated triode like the BJA requires four electrical connections: two for the filament, one for the grid, and one for the plate. Somehow, the electrodes inside of the tube must be brought to the outside world. The problem is more involved than simply drilling holes in the base, because it's aluminum and therefore conductive. The wires leading from the interior of the tube to the outside must pass through the aluminum base in an airtight fashion, while at the same time, remain electrically insulated from it. This is accomplished with a component I call a *feed-though*. Five feed-throughs are required for the BJA.

The BJA's feed-throughs were fashioned from brass, 1/8-inch NPT plugs. Using a 1/4-inch drill, I bored through the center of each plug. I cut a 1-1/2 inch length of brass tubing, and inserted it into the plug. The tubing was aligned so that one end was flush with the head,

Page 168

and the other end extended from the threaded end of the plug. With a torch and rosin core solder, the plug and brass tube were permanently bonded together. See figure 11.4.

The actual electrode terminals are composed of 1/16-inch brass rod or wire. Final assembly of the feed-through involves the combination of the modified plug, a 6-inch length of rod, and some liquid epoxy. It also helps to have a wad of modeling clay handy, as well as a plastic syringe.

I placed a fist-sized glob of modeling clay on my work surface, and inserted the rod so that it extended vertically. The modified plug was slid over it, threaded-side up, and pressed gently into the surface of the clay. The rod was then adjusted and aligned as necessary, to ensure that it lie at the precise center of the plug's bore. Remember: The wire must not touch the walls of the plug!

I mixed a batch of epoxy, and drew a small quantity of the adhesive into the syringe. I injected the epoxy into the space between the rod and the plug, filling it to its brim. Then, I left it undisturbed to set. See figure 11.5.

I fabricated five feed-throughs using this process. The threads of each feed-through were coated with teflon sealant before being screwed tightly into place. The heads of the feed-throughs lie on the exterior of the tube. The threaded end of the feed-throughs, with their projecting brass rods, extended into what will become the interior of the tube.

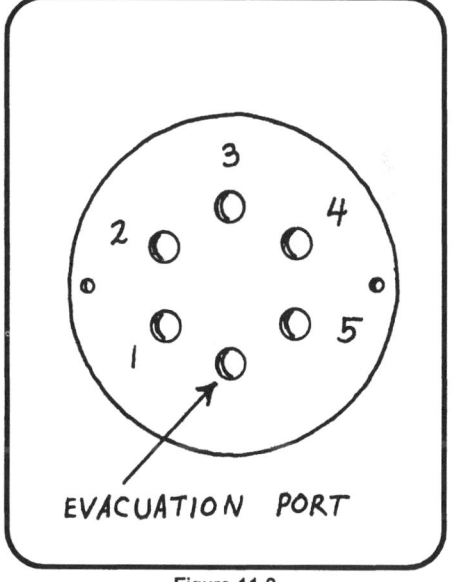

Figure 11.3
Hole pattern for the base of the Bell Jar Audion

I'd like to offer a few tips to help ensure success in fabricating feed-throughs. First, make certain all the constituent parts are meticulously clean and oil free. I like using isopropyl alcohol to cleanse them, though I've heard that lacquer thinner will do a good job, too. Once they're clean, try not to handle them with your bare fingers.

Page 169

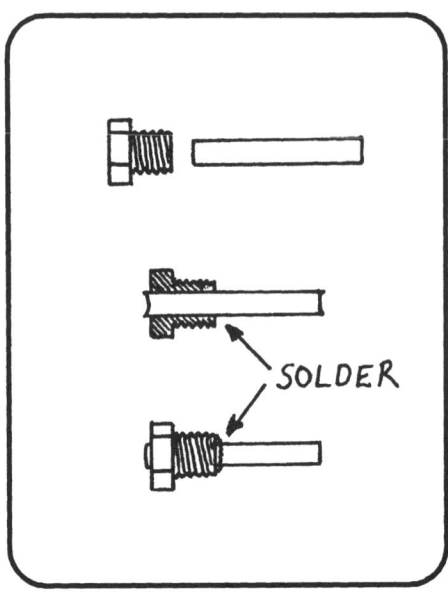

Figure 11.4
Assembly detail for feed-throughs

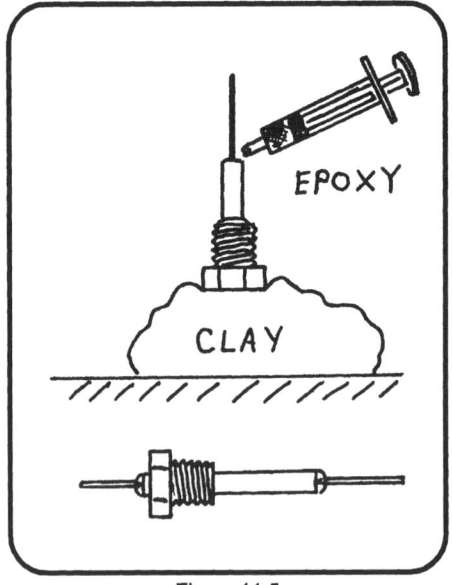

Figure 11.5
Filling feed-throughs with epoxy

Second, I suggest that the portion of the rod passing through the feed-through should be roughened to improve its bond with the epoxy. I used a pair of side-cutters to aggressively nick and notch the rod. Don't get carried away, though. If you nick the rod too deeply, it may be weakened and break off at a later time.

Finally, make sure to use a regular, slow-setting epoxy. In filling the feed-through, you will undoubtedly trap air bubbles. Regular epoxy, which stays fluid for a fairly long time, gives these bubbles ample opportunity to rise and pop. You can then use the syringe to add additional epoxy as necessary. Quick-setting epoxy, on the other hand, may harden well before these bubbles can escape, effectively trapping them. This defect leads to weak and potentially leaky feed-throughs.

The Grid

The first electrode I fabricated and installed in the triode was its grid. The BJA's grid is round, and measures about 2-1/4 inches in diameter. It consists of

Page 170

circular hoop or frame, strung with dozens of tiny grid wires. The grid wires span the hoop from left to right, in parallel, like the strings on a harp. The finished grid is reminiscent of a tiny circular barbeque grill.

The hoop was fashioned from 1/16-inch brass rod, which was formed around a suitable circular object. I believe I used an old bottle. After forming and trimming the hoop, the ends were brought together and joined with solder.

The grid wires themselves are composed of 28-gauge brass wire, commonly available at hardware and craft stores. Laying out the grid wires on the frame appears to be a daunting task at first glance, but it's actually quite easy if you know the trick.

I scrounged up an old corrugated cardboard box, and using a razor, cut a rectangle as wide as the hoop, but twice as long. The hoop was placed on the face of the cardboard, and centered.

I drew some 28-gauge grid wire off its spool and anchored the end to the cardboard with a piece of tape. From there, I began winding the cardboard as I

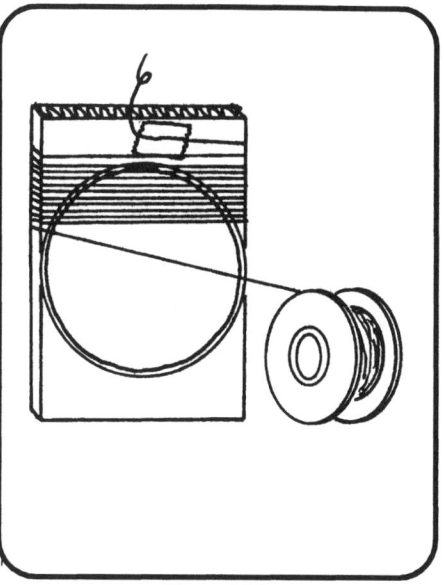

Figure 11.6
Procedure for winding grid wires

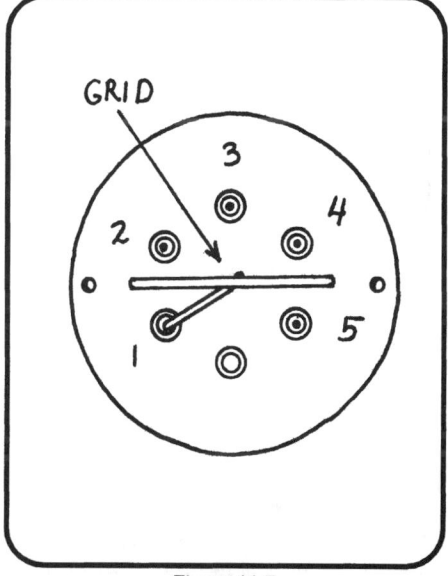

Figure 11.7
Locating the grid with respect to the hole pattern in the base

Page 171

would a tuning coil or other inductor. As the winding progressed along the cardboard form, I eventually reached the edge of the hoop. I continued to wind, up and over the hoop, so that I was now winding the wire around both the cardboard and the hoop. Throughout the process, I took special care to keep adjacent windings parallel, and evenly spaced. The gap from one winding to the next ran about 1/16 of an inch. Once I had wound my way from one end of the cardboard form to the other, I secured the final winding with another piece of tape. Figure 11.6 shows the grid winding operation.

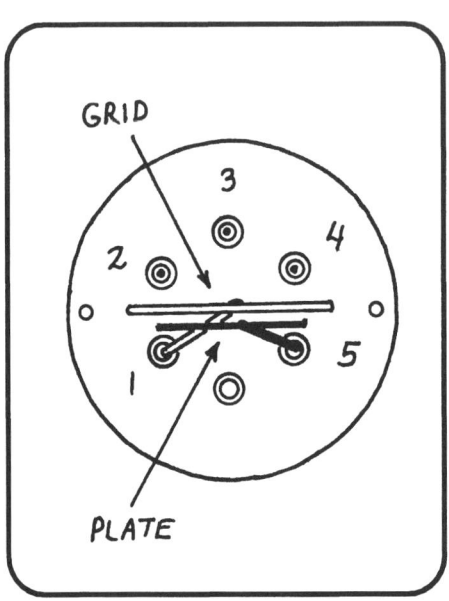

Figure 11.8
Locating the plate with respect to the hole pattern in the base

Next, using a small soldering iron and a light gauge of rosin core solder, I soldered all of the joints where grid wires came into contact with the hoop. Once all of the wires had been secured, it was a simple matter to cut the completed grid assembly away from the cardboard form and trim off all of the excess wire.

Installation of the completed grid involved selecting a feed-through to connect it to, and then soldering the grid into place.

Looking down upon the tube's base, with the feed-throughs and their rods pointing at me, I located the feed-through to the immediate left of the evacuation port. Starting there, and moving in a clockwise direction, I numbered the feed-throughs "1" through "5." In the BJA, the grid is connected to feed-through number "1."

The brass rods projecting from the feed-throughs are intended not only as electrical connections to the outside world, but also as the mechanical supports for tube's internals. Mechanically, the BJA's grid lies at the virtual center of the tube, so the rod sprouting from feed-through number "1" had to be bent, formed, and trimmed to hold the grid where it belongs. The grid assembly was soldered to the rod. See figure 11.7.

The Plate

The second and the simplest of the triode's three internal electrodes is the plate. The BJA's plate was fashioned from a disk of sheet brass, about 1-3/4 inches in diameter. In the BJA, it's connected to and supported by the brass rod extending from feed-through number "5."

Again, bending and adjustment of the feed-through rod is necessary. The plane of the plate should lie parallel to the plane of the grid, and the two should lie as closely together as possible. In the BJA, the gap from grid to plate was less than 1/8 of an inch. See figure 11.8.

The Filament

The original filament for the BJA was harvested from a 40-watt, tubular incandescent aquarium hood bulb. The bulb was cut open using the techniques described in the last chapter, and the long, thin, filament wire was extracted.

The BJA's filament was set up as an inverted "V". The rod projecting from feed-through number "3" was shaped like an inverted "L" with a shallow hook at its end. The filament was draped over and suspended by that hook. This point became the apex of the inverted "V". See figure 11.9.

The rod associated with feed-through number "2" was trimmed, and the end was tightly wrapped with a short length of 24-gauge copper wire. The free end of copper wire, in turn, was crimped onto one end of the filament wire. Feed-through number "4" was likewise modified and crimped onto the other end of the filament.

The Envelope

The envelope of the Bell Jar Audion was, of course, a bell jar. The jar I used measured about 3 inches in diameter and 4-1/4 inches in height. It was purchased at a craft store, and was intended as an enclosure in which to display heirlooms like pocket watches and such. Other glass containers of similar size and shape could easily be used

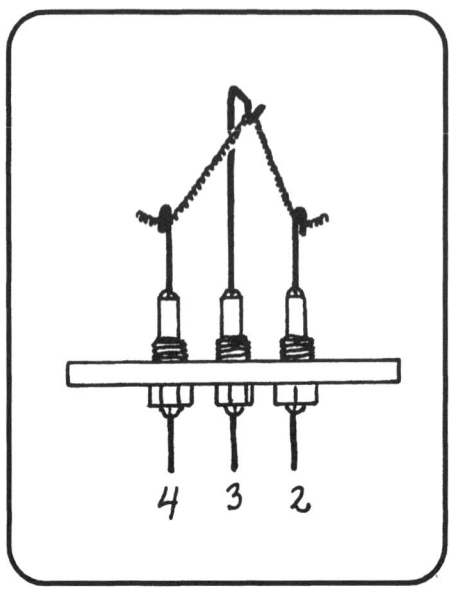

Figure 11.9
Details of the filament assembly

instead. I rather liked the bell jar because of its pleasing shape and optical properties.

The jar drapes over the electrode assemblies, and is sealed to the surface of the aluminum base with a generous, but neat, bead of black RTV silicone.

It's not enough, to simply locate the jar and then apply a bead of RTV along the seam. A ring of RTV should be laid down first, and then the jar should be set into it. I propose this, because the lip of the jar is somewhat irregular. A cushion of RTV between the glass and base plate not only effectively seals the joint, but also helps distribute the crushing forces that will be present when the jar is later evacuated.

The Stand and Terminals

The triode is functionally complete as described, but a tad unwieldy. The feed-throughs prevent the bulb assembly from sitting upright, and there's no convenient way to connect the electrodes to external equipment.

The Bell Jar Audion addresses these concerns with a permanent wooden stand composed of a circular wooden plaque, 7 inches in diameter, and two wooden thread spools, which serve as pedestals. Two 6-32 brass machine screws, 1-1/2 inches in length, were passed through the aluminum base of the tube, through the spools, and through the plaque. Washers and nuts secured them in countersunk holes from the bottom side of the plaque.

Connections are made to the tube through a series of binding posts or terminals comprised of 6-32 brass machine screws and matching brass, knurled, thumb nuts. 6-32 wing nuts would serve just as well.

If you look closely at figure 11.2, you'll note that each binding post features a brass label washer stamped with a letter. The letter helps identify the function of each connection— "P" for plate, "G" for grid, and so forth. The label washers were manufactured as follows.

I cut five brass disks from sheet stock, each measuring about 7/8 inch in diameter. I drilled a small hole in the center of each one, and then threaded them on to a 6-32 steel screw. I added a nut, and then tightened the stack.

Next, the assembly was chucked into an electric drill motor and spun. I played a metal file across the edge of the rotating stack, shaving off metal until the disks were uniform and roughly 3/4 inch in diameter. The disks were removed from the screw, polished, and then stamped with the appropriate letter. Letter stamps are available through tool distributors, either locally, or through mail-order catalogs. A set of stamps can cost as little as 10 dollars.

A short length of stranded, insulated wire was soldered to each feed-through, and threaded through a 5/8-inch hole bored through the plaque. Five shallow channels were cut into the bottom of the plaque using a router. The channels facilitate the passage of the each wire to its respective terminal post.

Once I had a feel for the fit and finish of the stand, I disassembled everything, stained the wood pieces, and sealed them with two coats of polyurethane. The Bell Jar Audion was then reassembled, and officially complete.

End Results And Going Further

I connected the BJA to my vacuum system and removed as much air as possible. In the beginning, it was quite difficult to draw much below 50 millitorr. At first I thought I had a significant leak, but as time wore on, the pressure slowly began to drop. It became apparent that the pump's initial inability to draw a harder vacuum had a lot to do with outgassing from the epoxy, silicone, and other materials used in the construction of the bulb.

I had no way of knowing if the evacuated bell jar could withstand the crushing forces of the external atmosphere, so I erred on the side of caution and covered the BJA with a clear plastic shield fashioned from a 2-liter soda bottle. It goes without saying that I wore appropriate eye protection.

Page 175

The internal pressure eventually dropped to, and stabilized at, 20 millitorr. At that point, I decided to light the filament.

Because of the length of the filament, and the nature of the lamp that it had been harvested from, I discovered that 100 volts was required to light it up. I had anticipated the need for a high-voltage supply for these types of experiments, and knocked one together from parts scrounged out of my junk box.

By all means, resist the temptation to power the filament through any circuit that ties directly to a wall socket. *This is extremely dangerous.* Even if you exercise care and caution, the sudden failure of any of the BJA's internal electrodes could result in a short circuit and your possible electrocution.

My power supply sports an isolation transformer as a highly desirable safety feature. High-voltage supplies like mine still pose a shock and burn hazard to the careless, but at least it protects me from direct contact with the supply mains. An overview of a homebrew power supply, suitable for use both as a high-voltage filament supply and an adjustable plate supply can be found in Appendix II.

I should note that when the filament was first lit, I observed a spike in the BJA's internal pressure. This was probably due to outgassing of the filament itself. While interesting, the effect was no big deal. A few minutes later, my vacuum pump had reduced the pressure to its previous level.

The performance of the BJA was measured with the test setup depicted in figure 11.10. A stack of 9-volt batteries, connected in series, was used as a "B" battery. The "B" battery was connected in series with a handheld digital multimeter set to measure current.

A similar battery stack, combined with a potentiometer, provided an adjustable grid voltage. A second multimeter was connected to the grid to measure the applied grid voltage.

The data table at the end of this chapter documents the electrical behavior of the first BJA, and clearly demonstrates amplifier behavior, namely, that changes in the grid voltage result in changes in the flow of current to the plate.

Encouraged by the first glance at the data, I hooked up a simple radio circuit, like that in figure 11.11. Subjectively, it appears that the BJA functions as a sensitive detector as well as an amplifier. I was surprised at the number of stations I could receive and their relative volume in the headphones, particularly when compared against a similar circuit using a crystal detector.

Initially, I was quite frustrated with the short life span of the BJA's filament. I was lucky to get 15 to 20 minutes out of any given tube. This isn't bad compared to De Forest's first Audion, which lasted a mere 3 minutes, but it made extended experimentation impossible.

The good news is that the silicone seal at the bottom of the bell jar is easily sliced with a razor. The bell jar lifts off, and the filament can be replaced. The bell jar is then reinstalled with a fresh bead of silicone, and you're back in business. Eventually, I got to the point that I could execute this sequence from start to finish, including the removal of the filament material from the donor bulb, in less than 30 minutes.

As I alluded to earlier, tungsten filaments are not very forgiving of manipulation. In an effort to prolong the life of my filaments, I began to consider how I might reduce the amount of handling required to transfer a filament from a lightbulb to my vacuum tube. My solution, I think, was rather novel.

If you've ever broken a lightbulb, you've seen the glass stem inside. Sealed into the stem are two wires that deliver power to the filament, and some additional wires that simply act as filament supports. I reasoned that if I could remove the stem and filament assembly as a complete unit, I wouldn't have to touch the delicate filament wire at all. I suspected that this might result in better filament life.

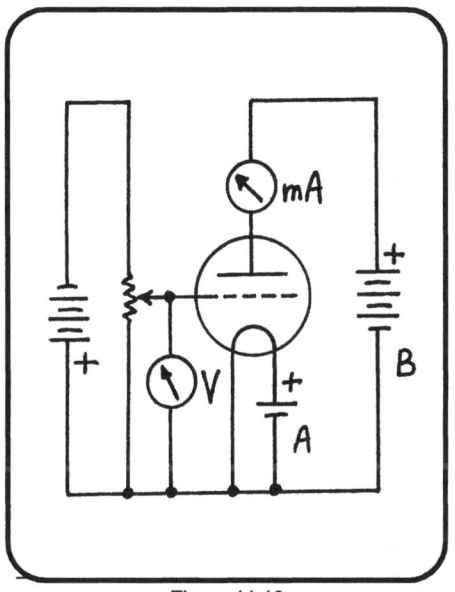

Figure 11.10
A test circuit for characterizing the behavior of the Bell Jar Audion

The extraction process is easier than one might at first imagine. Starting with an intact lightbulb, I scored the glass and removed the globe. Sometimes the bulb lifts off intact, other times it shatters, so it pays to wear gloves, goggles, and work over some kind of refuse container.

Next, I used a pair of slip-joint pliers to crush the aluminum base of the bulb. If this is done with patience and care, the stem and filament assembly generally breaks free with its lead wires still intact.

In most cases, you'll find a short length of thin glass tubing that runs up the center of the stem. This is the pipe through which the air was evacuated when the bulb was manufactured. I took advantage of this feature to mount the assembly in my Audion.

Using side cutters, I cut the "L"-shaped filament support off of feed through "3," leaving a stump pointing vertically. Next, I took the stem and slid the glass evacuation pipe over the rod. (If the tip of the evacuation tube is blocked off, nick the tube with small triangular file, and simply break the tip off.) The filament wires dangling from the stem were soldered to feed-throughs "2" and "4." That's all there is to it! Figure 11.12 shows the details of the new filament.

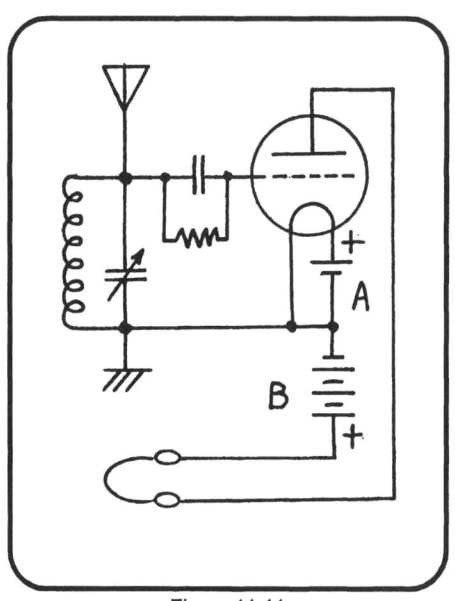

Figure 11.11
A simple radio circuit for use with the Bell Jar Audion

Following this revision to the BJA, filament life rose from minutes to *hours*. For that reason alone, I recommend this architecture over the inverted "V," previously described.

In addition, the ease of the transplant process made it possible to try filament material from smaller, lower-voltage bulbs. I tried a number of 12-volt automotive lamps, particularly those associated with brake lamps and turn signals. All of these exhibited excellent life, and the significant convenience of battery operation.

So, how does the Bell Jar Audion really perform? The inverted "V" filament didn't last long enough for me to develop a complete profile on the first BJA. I did do some scratch-pad calculations, however, and was disappointed to find the estimated voltage gain at something less than one.

What good is an amplifier whose output is less than its input? Don't forget the lessons learned with the Balance-Beam amplifier. The BJA's input impedance is very high, which means it poses an almost insignificant burden on the input signal source. The amplification subjectively apparent in my headphones may well have been due to a gain in power.

Strangely, I observed better performance in the BJA with filament assemblies taken from automotive bulbs. I'm not quite sure why this would be the case, though it may have something to do with the actual composition of the filaments themselves. Perhaps automotive bulbs, which are intended for use in an application where vibration is a factor, are composed of an alloy to make the filaments more durable. If so, the additives may incidentally improve the filaments' emission of electrons. In fact, at one point, I actually measured the voltage of an incoming audio signal versus the BJA's output, and saw the gain come very close to unity.

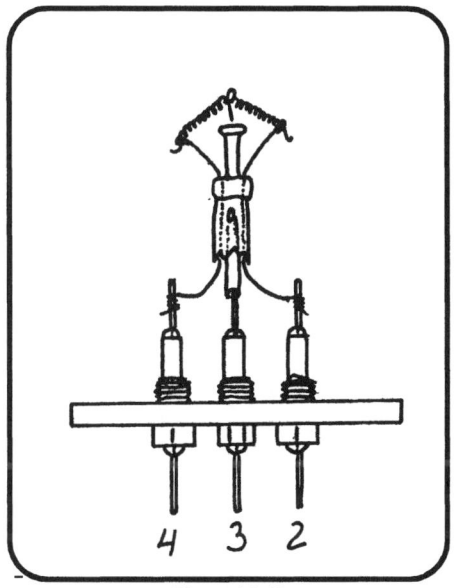

Figure 11.12
Using the glass filament assembly removed from a commercial incandescent light bulb

I can see no reason, in theory, why the BJA design couldn't be tweaked to deliver voltage gains greater than one, or possibly even two. The BJA is such a fun and convenient design that it lends itself well to modifications and improvements.

To begin with, there's much to be gained by reworking the grid to make its mesh more fine. I would also suggest adjusting the grid and plate to place them as close together as possible without allowing them to touch. As well, the filament can be positioned more closely to the grid, but be warned. The last thing you want is a hot grid. Finally,

Page 179

it occurs to me that there may be benefit in orienting the grid wires so that they run perpendicular to the axis of the glowing filament.

Of course, the vacuum system used to evacuate the bulb can always stand improvement. While the BJA is functional at the 20-millitorr level, sufficient gas remains in the chamber to limit plate voltages to something less than 60 volts. If you try to apply a plate voltage much higher than that, the residual gas becomes excited. The bell jar suddenly fills with a dull blue glow, and amplification behavior ceases. Pity on you if you're wearing headphones connected in the plate circuit. The glow is generally preceded by a loud hissing sound, followed, on occasion, by an ear-splitting shriek.

The more complete the vacuum inside the BJA, the fewer molecules remaining. Higher plate voltages can then be used to your benefit.

Bell Jar
Vacuum Triode

Pressure:	20 milliTorr
Filament Material:	Tungsten, removed from 120-volt, 40-watt tubular Edison-based lamp
Filament Voltage:	Approx 100 volts
Date:	8/5/00
Note:	Grid voltage applied with respect to negative filament terminal

Grid Voltage Applied	Plate Current (mA) With 45V Plate Voltage	Plate Current (mA) With 27V Plate Voltage	Plate Current (mA) With 18V Plate Voltage	Plate Current (mA) With 9V Plate Voltage
0.00	0.42	0.23	0.26	0.20
-7.50	0.41	0.22	0.23	0.19
-10.20	0.40	0.20	0.22	0.18
-14.10	0.39	0.20	0.20	0.17
-17.60	0.37	0.18	0.19	0.15
-21.00	0.35	0.17	0.17	0.13
-22.70	0.34	0.17	0.16	0.12
-26.00	0.33	0.16	0.15	0.10
-30.30	0.30	0.14	0.13	0.07
-34.50	0.28	0.12	0.11	0.04
-41.10	0.25	0.05	0.03	0.02
-46.70	0.22	0.03	0.02	0.02

References

Shunaman, Fred, "What Did DeForest Really Invent?" *Radio Electronics*, September 1961, Volume XXXXII, No. 9, Gernsback Publications, pp.˙ 47-49 (historical account of the invention of the Audion)

Chapter 12
Votive Triodes

The Bell Jar Audion is a wonderful first attempt at the construction of a thermionic triode. Its large dimensions and the easy accessibility of the internal components makes modifications and adjustments a snap. A patient tinkerer, willing to invest some time, can expect to get some decent performance out of that design.

That said, it doesn't take a great deal of reflection to identify a significant shortcoming in the planar electrode arrangement. Consider this: While the filament presumably emits electrons in all directions, the only ones that actually benefit the operation of the tube are those that head in the direction of the grid and plate. The rest, to a lesser or greater extent, are wasted.

If you take the trouble to inspect the interior of virtually any commercial vacuum tube made since the 1920's, you'll discover that the electrode arrangement is invariably coaxial. The electron emitter lies at the center of the tube, either in the form of a naked filament, or in the form of an indirectly heated cathode. A grid, implemented as a fine wire-mesh cylinder, surrounds the filament. The largest cylinder is solid and wraps around the other two. The latter, of course, comprises the plate.

With this architecture, the vast majority of the electrons emitted by the filament are essentially surrounded. Irrespective of the direction of their travel, sooner or later, their path will take them through the grid and to the plate. This represents a significant improvement in the utilization of electrons emitted from the filament, which translates to enhanced performance of the device.

In this chapter, we'll explore the designs for two experimental vacuum tubes with coaxial electrode arrangements. I was interested

Figure 12.1
The Tennis Ball Triode

in developing a technique for producing all-glass triodes without the need for glassblowing equipment. Both of these designs feature bases fashioned from glass ashtrays and glass bulbs that were originally intended for use as holders for common votive candles. For lack of a better name, I refer to them collectively as Votive Triodes.

The Tennis Ball Triode

The first Votive design I'd like to discuss is the Tennis Ball Triode. If you look at figure 12.1, you'll see that the tube's envelope is a small, clear, glass globe, which is reminiscent in my opinion, of a tennis ball. The globe measures about 4 inches in diameter and was purchased at a craft center.

The base of the TBT (Tennis Ball Triode) is an inverted glass ashtray measuring 4-1/4 inches in diameter. The particular ashtray I selected has two attributes recommending its use. First, it's quite thick— nearly 1/4 inch. This makes for a sturdy and stable base for the tube. Second, the bottom side of the ashtray features a round, shallow rim whose internal diameter just matches the outside diameter of the mouth of the glass globe. When the globe is inverted and placed upon the inverted ashtray, the ashtray's bottom rim and the mouth of the globe mate perfectly. This seam can be easily sealed with a thin bead of RTV silicone.

Incidentally, black and white photos fail to do this particular ashtray any justice. In reality, it's a deep, hypnotic, cobalt blue. The ashtray's color is what initially caught my eye when I first spied it on the shelf of a local factory-outlet store. Total cash outlay for the glass components ran about four dollars.

The TBT's Base

The bulk of the work associated with fabricating the TBT's base involves drilling numerous holes into the ashtray. All told, about 11 holes must be bored through its floor and sides. Before engaging in the drilling process, I first created a paper template to guide the operation. I began with a disk of paper sized to fit within the rim on the bottom of the ashtray.

I located the exact center of the disk, and using a common compass, I drew a series of concentric circles. The circles had a radius of 3/8 inch, 11/16 inch, and 1 inch. These define the domain of the filament, grid, and plate, respectively. Note that there's nothing magical about these values. I specify them merely as a record of the internal dimensions of my TBT. They allow for liberal distances between the TBT's internal parts, which makes assembly and experimentation more convenient. Feel free to change them as your raw materials and skill allows. As we've seen in previous chapters, there are

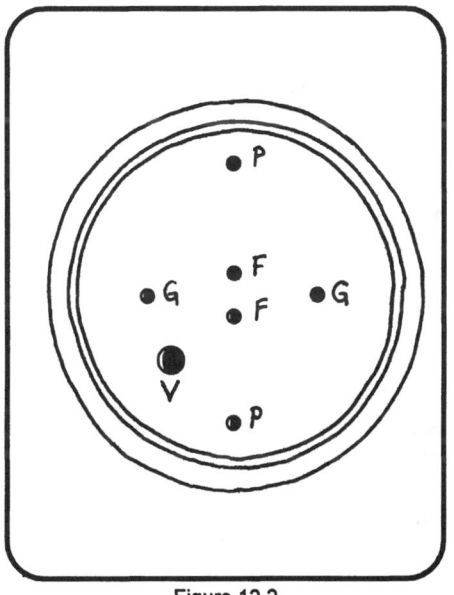

Figure 12.2
Location of holes in the Tennis Ball Triode's base

worthwhile performance gains to be had in closing up the space between adjacent electrodes.

With circles laid out, the next step was to locate the appropriate positions for holes. Figure 12.2 is a map of that hole pattern. Two holes are required to support the filament structure. One is located at the exact center of the paper disk, the other lies on the first circle at the 0-degree position. Two more holes are needed for the grid assembly. They're located on the second concentric circle at the 90-degree and 270-degree positions. The plate assembly requires two holes on the third circle at the 0- and 180-degree positions. Last but not least, a hole is needed for an evacuation port. This hole also lies on the third circle, but is located at the 135-degree position. As I determined the placement for each hole, I marked its located with a small "+."

The template was afixed to the bottom of the ashtray with rubber cement. Using a 1/8-inch glass drilling bit, I bored all of the required holes right through the template. I switched to a 1/4 bit, and enlarged the exhaust port.

Four additional holes must be bored into the wall of the ashtray to facilitate the installation of connection terminals. Using a narrow strip of paper as a template, I laid out the pattern. My ashtray features four notches in which a smoker may rest his cigarette. Taking this into account, I spaced the holes evenly between successive notches, set back about 1/2-inch from the open mouth of the ashtray. The pattern was fixed to the glass with rubber cement, and the 1/8-inch holes were drilled right through the paper.

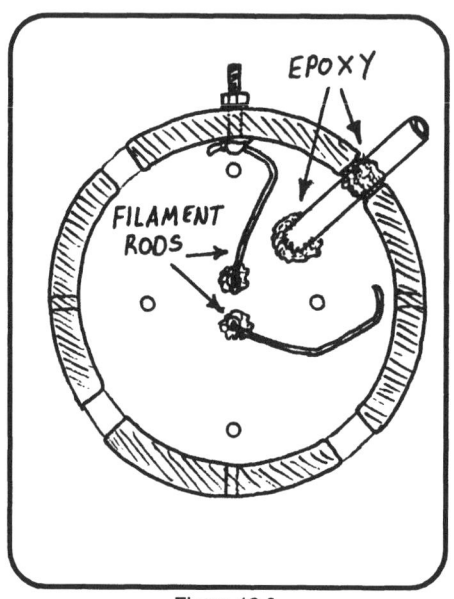

Figure 12.3
Details of the filament leads and the exhaust port

With the drilling operation complete, I removed the templates and thoroughly cleaned the glass with soap and hot water. The objective here was to remove all traces of grease or oil.

Four, 1-inch, 6-32 brass screws were used as terminals in the TBT. They were inserted through the holes drilled into the wall of the ashtray and project from the inside out. They were each secured with a brass nut, and then fitted with a knurled thumb nut. Great care must be exercised in tightening the nuts on these screws, because the wall of the ashtray is curved. If too much torque is applied, the ashtray is liable to be broken. I found that it helps a great deal to use small cardboard or fiber washers wherever a screw head or nut contacts the glass. Fiber washers yield to pressure and help to distribute the forces on the surface of the glass.

The evacuation tube for the TBT consists of 6 inches of 1/4-inch soft copper tubing. The tubing must be bent into an "L"-shape. This can be difficult if attempted freehand, because metal tubing has an annoying tendency to kink. Tubing benders are cheap and available at hardware stores. A possible alternative is to rest the copper tubing between the flanges of a small pulley, and use that as a guide to impart a graceful arc as it's bent.

With the ashtray inverted, the foot of the "L" was inserted through the evacuation port from below. The long end of the "L" was manipulated to project from the side of the ashtray. It passes through one of the cigarette slots in the wall of the ashtray.

The evacuation tube was cemented into place with appropriate amounts of quick-setting, gel-type epoxy. The epoxy was applied to the tubing, and the tubing was slowly twisted to-and-fro as it was inserted into the evacuation port. This action is intended to help distribute the epoxy in the gap between the metal and the glass. It's okay if a small glob of epoxy collects on the interior of the ashtray as the tubing moves into position. It can only help the seal. I also applied a liberal gob of epoxy to the tubing where it rests in the cigarette slot. This gives the tube mechanical stability, and helps take strain off of the exhaust port's glass-to-metal seal. See figure 12.3.

The TBT's Filament

Two 1/16-inch brass rods carry filament power into the TBT. Installation was a trivial exercise.

The first rod was inserted through the center hole, previously bored through the ashtray's bottom. It was adjusted to project about 1-

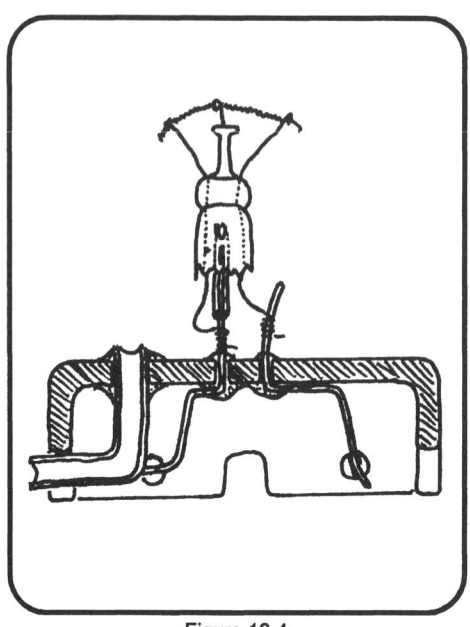

Figure 12.4
Cutaway of the base, showing the exhaust tube, filament support wires, and details of the installation of the filament assembly

1/2 inches into what will become the interior of the triode.

Inside the ashtray, the rod was shaped, trimmed, and bent so that it traveled to one of the terminal screws. The end of the rod was set into the slot in the screw's head and soldered into place. The second rod was installed by the same means. It, of course, was routed to a different terminal screw, and likewise soldered into place.

Once again, I mixed up a small batch of gel-type epoxy and used it to create a seal where the brass rods pass through the floor of the ashtray.

The TBT's filament structure is based on the lessons learned from the Bell Jar Audion. I gave up trying to make my own filaments, opting instead to use the intact stem assemblies harvested from some automobile tail lamp bulbs.

Once a stem assembly has been extracted from an appropriate donor bulb, installation in the TBT is easy. The stem is slid onto the center filament rod. One wire, dangling from the stem, is soldered to the center rod. The other wire is soldered to the second run. Note that one or both of the brass rods may need some trimming to accommodate the stem assembly you're using. See figure 12.4.

The TBT's Grid

The TBT's grid is a cylindrical wire mesh cage. The mesh is a tight spiral of 28-gauge brass wire wound around eight vertical ribs composed of 1/16-inch brass rod. Fabricating the grid is not overly difficult if you know the trick.

I began by locating a suitable cylindrical form on which to wind the grid. In the case of the TBT, a dead "D" battery was the perfect size. A small bottle or a piece of pipe would probably have worked just as well.

I cut eight pieces of 1/16-inch brass rod, each 3 inches in length, to use as ribs for the grid cage. Using bits of electrical tape to hold them in place as I worked, I secured the ribs to the case of the battery. The ribs were oriented to run parallel to the length of the battery, and arranged so that they were spaced equally around its circumference. A band of tape wrapped around each end of the battery kept the ribs in firm alignment.

Next, the cage was carefully wrapped with the 28-gauge grid wire. I drew some wire off the supply spool, anchored the end to the battery with a bit of tape, and began the winding operation. I tried to maintain a uniform spacing between adjacent turns with as fine a gap as I could reliably produce. I was able to achieve grid spacing better than 1/16 inch. I continued winding until I had produced a cage roughly 2 inches in height. I clipped the wire off the supply spool, and secured the loose end to the battery with more tape. See figure 12.5.

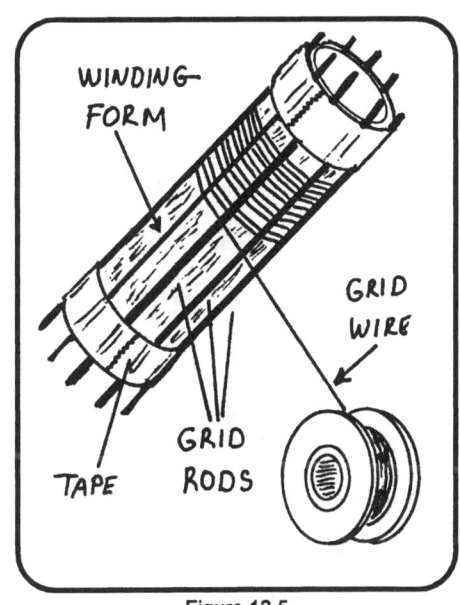

Figure 12.5
Fabrication of the cylindrical grid assembly

I made some minor adjustments to the grid wires, using a toothpick as a tool. When I was satisfied that all was right, I began soldering the intersections between grid wire and rib. When the soldering was complete, I was able to remove all of the electrical tape and slide the completed grid off the battery.

I used side cutters to trim the ribs to their proper length. In fact, all of the ribs were cut down to match the height of the cage except for two. These would become the legs on which the grid stands.

I considered inserting those legs through the appropriate holes in the ashtray and sealing them with epoxy. Unfortunately, there's a catch. Based on my experience with the Bell Jar Audion, I expected the TBT's filament to burn out sooner or later, and I'd want to be able to replace it. If you think about it for a moment, it would be next to impossible to work on the filament if it were surrounded by the grid and plate. I decided that both of the latter elements would have to be removable.

My solution to the problem was to employ binding posts. Two small binding posts were fastened to the ashtray with 6-32 screws. The screws, the grid holes bored through the ashtray, and the base of each binding post were gobbed with epoxy so that, when the screws were tightened, they formed an airtight metal-to-glass seal.

If you look at figure 12.6, you'll note that the end of each leg of the grid assembly is terminated with a sharp, "L." The foot of each "L" is inserted and secured in its respective binding post. With this arrangement, the grid can be removed and reinstalled as many times as you see fit.

Figure 12.6
Binding posts, used as anchors for the grid and plate, allow for later removal

I arbitrarily choose one of the two binding posts to serve as the electrical connection to the grid. I soldered a short piece of bare wire to the head of the screw that anchors one of the grid's binding posts to the glass. I routed that wire to the wall of the ashtray, and soldered it to the head of one of the unused terminal screws.

The TBT's Plate

The plate of the TBT is a basic cylinder, 2-1/4 inches in diameter, and 1-1/2 inches in height. It was fabricated from a strip of brass shim stock. Two sections of 1/16-inch brass rod act as legs for the plate.

Like the grid, the plate's legs are terminated with "L"-shaped feet. The feet are inserted into a second set of binding posts. I soldered a short wire to one of the screw heads that anchors the plate's binding posts. The wire was routed to the remaining terminal screw where it was soldered into the slot in the terminal screw's head. See figure 12.7.

The TBT Completed

To complete the assembly of the Tennis Ball Triode, I gave the internal electrodes a final once-over, ran a bead of black RTV silicone around the ashtray's aforementioned rim, and then carefully lowered the glass globe to meet it. I left the device undisturbed for 24 hours. I always wait at least that amount of time to ensure the silicone has cured before applying a vacuum.

I inverted the finished TBT, and peered into the bowl of the ashtray, wondering what I could do to enhance or strengthen the fragile epoxy seals. I was also concerned that any undue manipulation of the copper evacuation line might fracture its seal as well. It occurred to me that I could fill it with liquid plastic, which I presumed would bond everything into a solid, airtight block. I mixed up a batch of clear casting resin of the kind

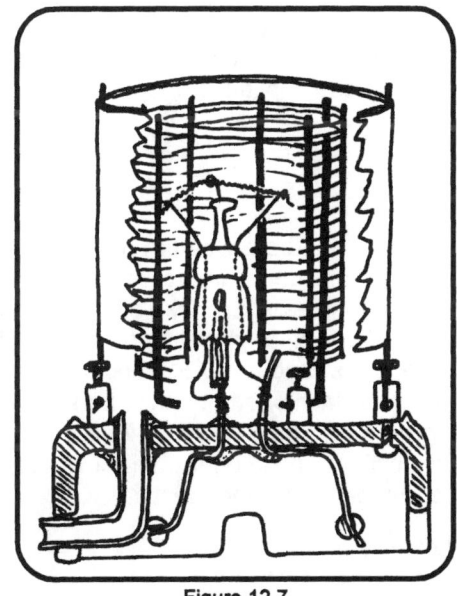

Figure 12.7
A phantom view of the completed Tennis Ball Triode, without the its glass globe

commonly used to make paperweights, and dumped it in.

What I failed to recognize was the fact that a good casting resin, by design, should shrink slightly on setting. This is to facilitate the removal of the finished article from its mold. Predictably, the resin in the ashtray shrunk slightly on curing, which caused it to draw away from the glass. The resulting gaps defeated the whole purpose of pouring resin into the base of the triode. The only positive thing that could be said about the casting process was that the evacuation tube was very much strengthened and stabilized.

The Tennis Ball Triode, thus completed, sat nicely on my bench top and could easily have been used in its raw form. I opted instead to mount it to a rectangular, stained wooden plaque, measuring 7 by 9 inches. The TBT was secured to the plaque with 6-32 brass machine screws and a set of four brass angles. The angles also act as conductors to bring the triode's various signals to the underside of the plaque.

As with the Bell Jar Audion, I fabricated a set of brass terminals with label washers and installed them along the front edge of the plaque. Shallow channels, cut into the underside of the plaque, allow wires to run between the filament, grid, and plate terminals at the front of the plaque to the corresponding TBT terminals.

I also fashioned a small metal bracket to anchor and support the copper evacuation tube. The bracket was fixed to the plaque with 6-32 screws and nuts. The bracket bears the brunt of torque, or any other forces applied to copper tube when the TBT is in use.

The Beehive Triode

I had barely begun playing with the Tennis Ball Triode when I began considering improvements to the design. I had another ashtray handy and decided to take another stab at tube building. In this case, instead of a clear globe, I would house the tube's internals in a deep, blue, glass candle cup. Resting on the ashtray, its shape is reminiscent of an old-fashioned beehive, hence the name "Beehive Triode." See figure 12.8. The Beehive Triode, or BT, is neat to look at, particularly when it's lit and in operation. However, its distinguishing features run far deeper than mere aesthetics.

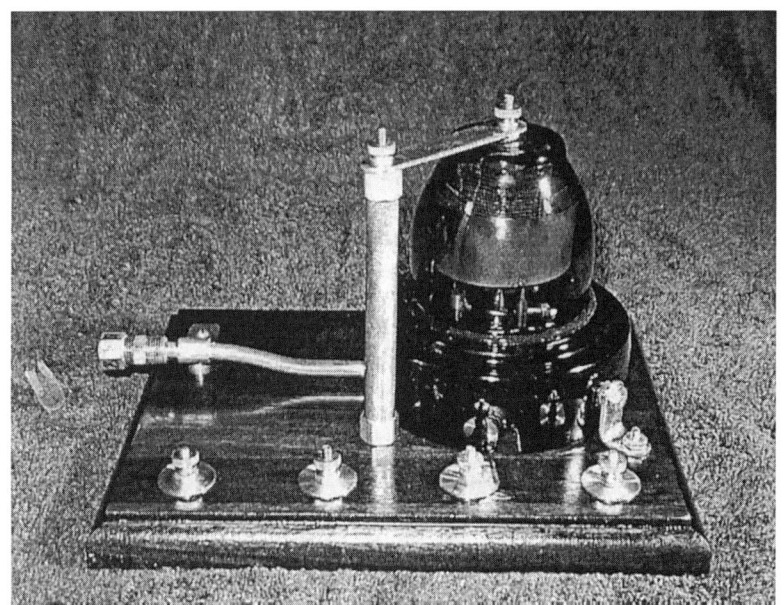

Figure 12.8
The Beehive Triode

BT Binding Posts

Among several things that annoyed me about the Tennis Ball Triode were the binding posts that secured the grid and plate to the base. While they definitely simplify removal and facilitate the replacement of a burned-out filament, I found them to be cumbersome. For one thing, you can't really grab the thumbscrews with your fingers, because the plate and grid lie in the way. Your only alternative is to use needlenose pliers to reach in from the side and loosen them.

Once the binding posts have been loosened, the plate and grid must be rotated in order to pull their "L"-shaped feet out of the binding posts. Only then can they be lifted up and away. Complicated procedures like this expose them to stress and inadvertent damage.

Another thing that bothered me was the relative availability of the binding posts themselves. I purchased them from the catalog of a mail-order vendor who specializes in antique radio supplies. The last time I checked, they were still available, but I hate designs that depend too heavily on unusual, single-source components.

My answer was to develop my own binding post, which appears in figure 12.9. Instead of a hole in its side and a screw on top, my design features a hole in the top, and a screw on the side. The benefit of this reorientation should be obvious. The screw is always accessible with a screwdriver, regardless of its placement in the triode. As well, the legs supporting the tube's internal assembles no longer need "feet." Once the screw has been loosened, a plate assembly, for example, can be lifted straight up and out without additional jockeying to get it clear of the posts.

The BT's custom binding posts were fabricated from 1/4-inch brass rod. To fashion a binding post, I began by cutting a 1-inch length of the rod. Using my lathe, I trued up and squared the ends.

I arbitrarily chose one end of the rod to be the top end. I located the exact center of the top face and began drilling vertically through the axis of the rod using a 1/16-inch bit. I stopped when I had reached a depth of 3/8 inches.

Next, I drilled a horizontal hole into the side of the rod using a #43 tap drill. I penetrated just far enough to meet the vertical passage bored out in the previous step. The side hole was tapped and fitted with a short 4-40 machine screw. Thread taps and appropriate tap drills are available at hardware stores.

As a final step, the diameter of the bottom half of the binding post must be reduced. This was easily done on my lathe, though the same effect could be achieved by chucking the binding post in an electric drill and dressing it with a metal file. In the end, the top 1/2 inch of the post measures 1/4 inch in diameter. The bottom diameter measures 1/8 inch.

To install these posts into the ashtray, I simply epoxied them into their respective holes. If you wanted to get fancy, you could probably use a 6-32 die to cut threads into the narrow portion of the binding post, wet it with epoxy, slide it through the appropriate hole in the ashtray, and secure it from the other side with a nut.

The BT's Grid

Another attribute of the Tennis Ball Triode that I found unattractive is the relative complexity of the tube's base. You need to drill quite a few holes to accommodate all of the rods and screws passing through the glass. Bear in mind that each hole represents an opportunity for leakage. This complexity also implies a limit to how small and compact the electrode arrangement can be made. At the heart of the problem is the fact that all of the tube's internal structures are mounted to, and supported by, the same surface.

The electrode arrangement in the Beehive Triode represents something of a departure from this tradition. It's still a coaxial triode, with a filament surrounded by a cylindrical grid and plate, but the grid is not secured to the ashtray at all. In fact, it's fastened to the tube's envelope! When the BT's glass candle cup is lifted off the triode, the grid goes with it, leaving behind the filament and plate still affixed to the surface of the ashtray.

There are several reasons why this mounting scheme represents an improvement. For one thing, three fewer holes need to be drilled in the ashtray, and the mechanical clutter on the surface of the ashtray is significantly reduced. The replacement of a blown filament is greatly simplified. Only the plate assembly needs to be removed, because the grid ceases to be an impediment the moment the candle cup is removed. The grid, by the way, need never be touched. It remains permanently attached within the candle cup, safe from accidental damage.

The construction of the BT is very similar to that of the Tennis Ball Triode, with the exceptions that the improved binding posts are used, and only two are needed (to support the plate) as opposed to four.

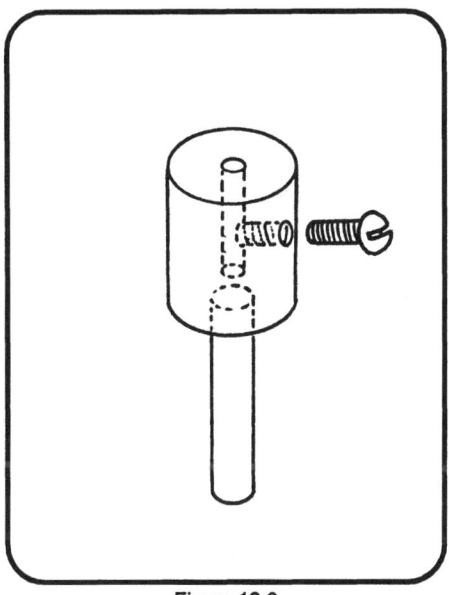

Figure 12.9
Details of the Beehive Triode's custom binding posts

The BT's grid assembly is similar to the TBT's, and is fashioned by the same process. The BT's grid, however, has a cap or lid. The lid is a small brass disk, cut to the same diameter as the grid cage. It's soldered into place.

The peak of the Beehive Triode's envelope, that is to say, the base of the inverted candle cup, must be drilled with a single, 1/8-inch hole. A 6-32 threaded brass rod is inserted through the hole and secured on either side with brass nuts. It goes without saying that the rod's threads are well saturated with epoxy where the rod passes through the glass.

The grid assembly is slid onto the threaded stub projecting into the tube, and secured with a brass nut. On the outside of the tube a knurled brass thumb nut serves as a terminal. See figure 12.10.

The completed Beehive Triode was closed up and sealed with a bead of black RTV silicone rubber. I revisited the idea of pouring plastic into the base of the tube, only this time, I arrived at a suitable material—liquid epoxy. Unlike the casting resin, it wetted and remained stuck to all of the glass surfaces, and hardened without contracting.

Like the Tennis Ball Triode, the BT was affixed to a rectangular wooden plaque with a set of brass "L"-brackets. Four brass screws with thumb nuts serve as terminals. Short insulated wires connected to those terminals carry signals through channels routed into the bottom face of the plaque, where they finally connect to the terminals on the tube itself.

The metal structure bolted to the tip of the tube, visible in figure 12.8, is merely the grid connection to the tube. The vertical column consists of a 4-inch section of brass tubing, capped at either end with a brass compression nut. The nuts aren't threaded onto the tube, they're just pressed into place. The column is held erect, and at the same time, anchored to the plaque with a length of 6-32 threaded rod. The rod, fitted with a suitable nut, passes through the plaque, up the center of the column, and through the top, where it's terminated with a brass thumb nut.

The horizontal component is a 3-inch length of brass strip, 1/2 inch wide and 1/16 inch thick. It bridges the span from the grid terminal on the top of the tube to the brass column described above.

End Results And Going Further

Both the Tennis Ball and Beehive Triodes demonstrated obvious improvement over the Bell Jar Audion. In fact, the latter's performance eventually rose to a level on par with some of the earliest commercial tubes.

I began subjective testing by hooking each tube to a series of simple tuned radio circuits. Both of the Votive Triodes performed well as radio detectors. In addition, there was ample evidence of amplification. Musical content in the receiver's headphones was noticeably louder using one of the triodes, when compared to an identical circuit using only a germanium (diode) detector.

One version of the blue Beehive Triode exhibited enough gain to function in a primitive regenerative radio circuit. See figure 12.11. A regenerative circuit is one in which a portion of the amplifier's output is purposefully fed back to its input. This feedback loop, if set up properly, recirculates the signal and affords the amplifier (in this case, my triode) the opportunity to build upon successive iterations of its own amplification. A comparatively poor triode, used in a circuit of this type, can be made to perform like a tube that's many times better.

I eventually took the trouble to try to characterize the tubes using the same equipment and technique described in the Bell Jar Audion chapter. An example data sheet for the Tennis Ball Triode appears at the end of this chapter.

If you compare the response of the Tennis Ball Triode to the Bell Jar Audion, it's easy to see that the coaxial tubes not only carry larger currents, but are also capable of dramatic changes in those currents through the application of much smaller grid voltages.

For example, using a 45-volt plate supply, the

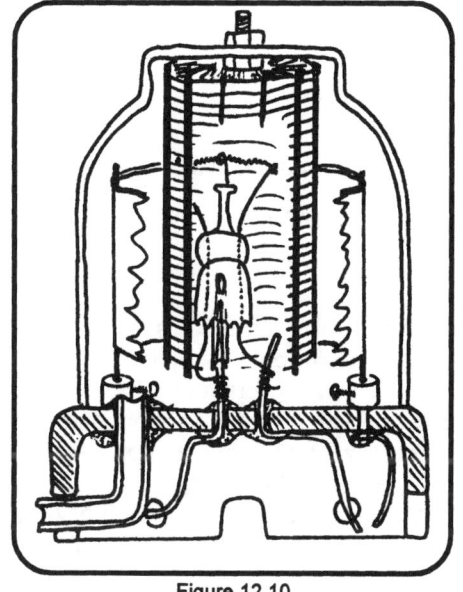

Figure 12.10
A phantom view of the completed Beehive Triode, with its glass envelope

maximum current carried by the Bell Jar Audion was 0.42 milliamps. The Tennis Ball Triode, under identical conditions, was able to carry 1.06 milliamps. That's two-and-a-half times better!

In order to throttle that current down to zero, the Bell Jar Audion required a signal greater than 50 volts applied to the grid. The Tennis Ball Triode, on the other hand, was able to bring the plate current to zero with as little as 6 volts applied to the grid. In other words, the Votive Triodes were more effective at controlling the local

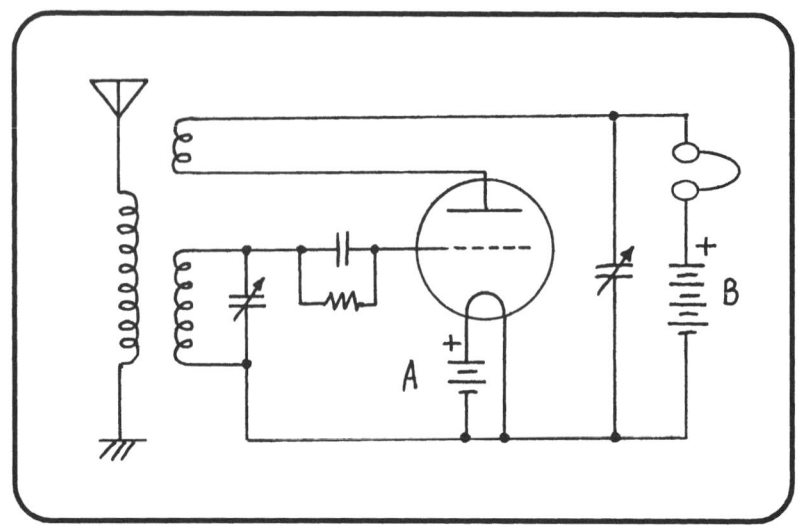

Figure 12.11
A simple, yet sensitive regenerative radio circuit used successfully
with the triodes in this chapter

(plate) supply on the basis of a remote input signal (to the grid). Put simply, they're better amplifiers.

How do you determine a homebrew tube's gain? Computing gain on the basis of data like those at the end of the chapter can be a little tricky. Part of the reason has to do with the fact that a varying *voltage* is fed into the tube, but the amplified result is a varying *current*. Simply dividing one into the other may not help you arrive at a gain value, because in essence, you're comparing apples to oranges. (Note: The ratio of plate current to grid voltage is a value that actually *does* find use in the design of amplifiers. Vacuum tube engineers call it *mutual conductance*. It's possible to compute the voltage gain of the tube if you know the specific load it'll be driving, the internal impedance of the tube, and its mutual conductance.)

Since we'd like, ideally, to compare apples to apples, one approach to quantifying a tube's amplification potential is to carefully select input and output parameters so that we are, in fact, comparing like measurements.

Take a look at the Tennis Ball Triode data at the end of this chapter. A quick glance suggests that there is more than one way to change the plate current flowing in the tube.

For example, if the grid is kept at a fixed value, but we reduce the plate voltage, the plate current goes down. Likewise, if we keep the plate voltage constant, but apply a grid voltage, the plate current also goes down. Stated another way, there are two factors, internal to tube, that influence the degree to which the plate current is varied. Note that one is an input factor associated with the grid's ability to diminish the plate current, and other is an output factor associated with the plate's tendency to diminish the plate current as the plate voltage is reduced. If we divide one into the other, we're left with a value engineers call the *amplification factor* of that tube.

Numerically, the amplification factor equals the plate-current change produced by a given grid-potential change, divided by the plate current change produced by an equal plate-voltage change.

Referring back to the data table, we see a 1.68-milliamp reduction in plate current if we change the grid potential by 5 volts. On the other hand, if we vary the plate voltage by 5 volts, the plate-current changes by 0.7 milliamps. If you take the 1.68 and divide it by 0.7, you get a resulting amplification factor of 2.4

The easiest way to measure a homebrew tube's gain is to connect the tube to a real load, and use the familiar out-divided-by-in relationship to calculate a final value. On one occasion, for example, I applied a test signal to the grid of the Beehive Triode and measured it with an oscilloscope. Next, I measured the magnitude of the signal being delivered to my headphones. I divided the measured output by the input, and computed a voltage gain to be greater than 3.

Filament life is still an issue with these tubes, and the topic bears some additional discussion. For Votive Triode filaments, I made exclusive use of the stem assemblies from 12-volt automotive-type lightbulbs. By and large, I prefer tail-lamp bulbs, as they tend to have thicker, more robust filaments that end up lasting longer.

I also tinkered with a class of bulbs intended for use in motor homes and recreational vehicles. RV bulbs look, for all intent and purposes, to be common, 110-volt lamps, complete with a standard Edison screw-type base. The filaments, however, are 12 volts! While the bulbs are more expensive to dismantle and destroy than tail-lamp bulbs, their stem assemblies and filament strands are much larger, and consequently, easier to work with.

I never lit my filaments until I was certain that all outgassing had occurred, and the tube's internal pressure was as low as it was going to get. My best repeatable pressures lie in the 15- to 20-millitorr range. In addition to improved triode performance, harder vacuums seem to prolong filament life.

There is also a very definite link between a tube's filament life and how hard it's been driven. An interesting 1922 reference to the subject states, "*... it can be shown that to double the filament emission will reduce the operating life of the tube to one-fourth, whereas, by operating the filament at 95% of its rated voltage, the life will be doubled.*" In other words, to prolong the life of your tubes, keep your filament at the lowest useable voltage at all times.

Initially, I'd power all of my filaments with half their rated voltage. This level of heating usually produced useable electron emission. As time wore on, however, I observed a gradual degradation in emission. Incrementally higher voltages were required to maintain the emission values measured at the onset. At the end, I generally had to drive the filaments to 120% of their rated voltage to get them to work. Shortly thereafter, they'd develop hot spots and disintegrate.

The business of filaments and useable emission got me to thinking about thoriated filaments, and how they might be manufactured on an experimental basis. It occurred to me that the coatings on (Welsbach) gas mantles contain both thorium and cesium oxides, both of which, if successfully extracted, could be applied by some means to my tungsten filaments. Under the action of a vacuum and the heat of a lit filament, the thorium oxide might be partially reduced to yield trace amounts of metallic thorium.

I must warn you that, while interesting, this idea is not only untested, but possibly dangerous. To begin with, thorium is mildly radioactive, as are some of its decomposition products. Crushing a mantle to extract thorium compounds implies the production of dust, a material you simply do not want to inhale or ingest. In fact, concerns for this have motivated industry to develop mantles composed of other, less harmful substances.

Another problem has to do with binding agents. How would you get the thorium-oxide powder to stick to the filament? Moisture attacks metallic thorium, I'm not certain what it does to its oxide. Other binding agents could be used, but under the action of a white-hot filament, would they decompose into substances that would act against the thorium?

In practice, thorium is probably not the sort of thing you'd want to mess with. On the other hand, there's nothing to stop you from experimenting with filament coatings composed of less malignant substances.

Coaxial triodes like the Tennis Ball and Bee Hive triodes represent interesting variations in the standard filament-grid-plate arrangement. As a parting shot, let's consider an electrode geometry

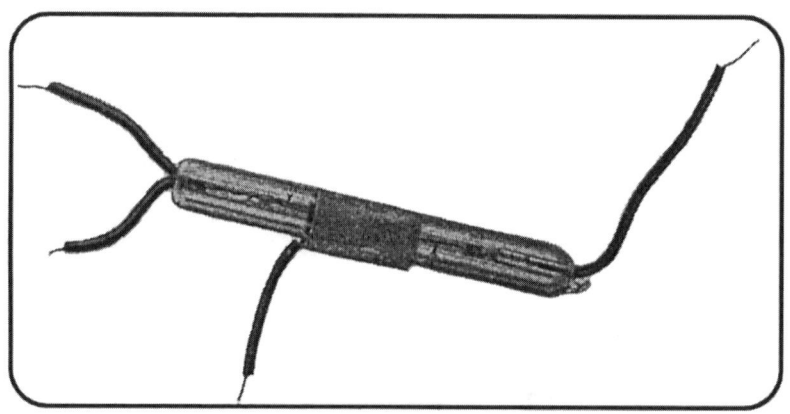

Figure 12.12
An example of a genuine Weagant tube
(Adapted from Vacuum Tubes in Wireless Communication, 1918)

that is, if nothing else, radical. How about a triode with *no* grid at all? Back around 1913, the chief engineer of the Marconi Wireless Telegraph Company of America introduced what was then referred to as the *Weagant Oscillation Valve*. An image of the tube appears in figure 12.12.

Like the other triodes we've constructed, the Weagant tube contains a heated filament to emit electrons, and a charged plate electrode to collect them. The difference lies with its grid, or rather, the complete absence of one. The Weagant tube does not rely on an internal electrode to regulate electron flow, but rather, an *external* one!

The tube's body is long and narrow. Wrapped around its waist is a cylinder of copper foil. Applying an electric charge to that foil creates an electric field in the interior of the tube. That field interacts with the electrons streaming from the filament to the plate. Reportedly, it's as capable of regulating the plate current as a grid of the common variety.

I was quite surprised to stumble upon technical references to the Weagant tube as I'd never heard of it before. It obviously failed to be a commercial success, and I have to presume there were probably good reasons for it.

That said, the Weagant architecture has several stated advantages. For one thing, it's structurally simpler than the other triodes we've visited. There's one less electrode that needs to be sealed in the glass, and none of the remaining electrodes require elaborate support structures. Spacing of the electrodes is said to be

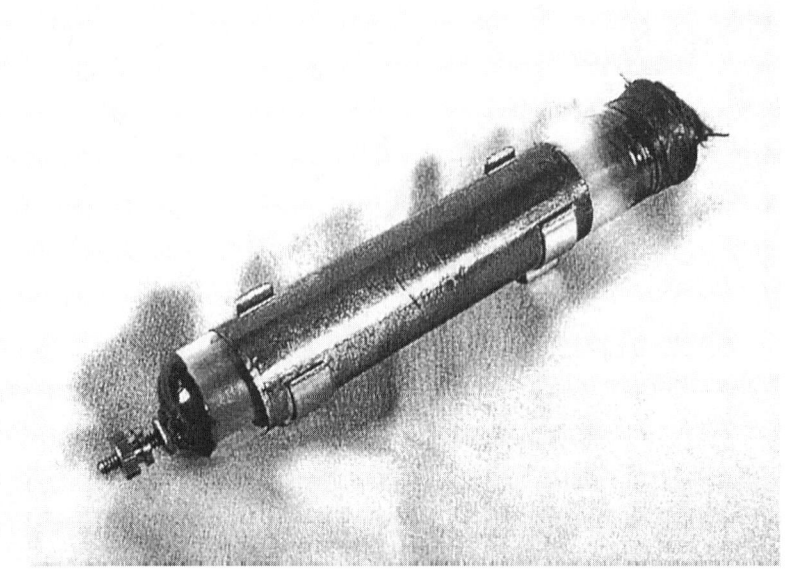

Figure 12.13
The Hamster Bottle Triode

non-critical. In fact, they need not be placed closely together at all. Finally, the Weagant tube is alleged to function in any of the common amplification or receiver circuits of the day. For the vacuum tube enthusiast, The Weagant tube offers an irresistible opportunity for experimentation.

Yes, I took a stab at building one myself, a model I call the Hamster Bottle Triode. See figure 12.13. The body of the tube is a medium, clear, glass test-tube taken from a water dispenser used for small caged animals like gerbils or hamsters.

Using a propane torch, I heated the sealed end of the hamster bottle and used a wire to poke a hole through the glass. I continued to apply heat until the hole enlarged itself, and its edges were fire polished. Once the tube had cooled, I installed the plate.

The Hamster Bottle Triode's plate is a 1/2-inch copper pipe cap soldered to the head of a 6-32 brass screw. See figure 12.14. The threads of the screw and the top of the cap were liberally coated with red high-temperature RTV, and inserted into the tube. The shaft of the screw was fed through the new hole in the sealed end of the tube, and secured on the outside with a fiber washer and a nut.

Admittedly, the original HBT (Hamster Bottle Triode) was fashioned before I had developed my technique for removing the stem assemblies from lightbulbs. Consequently, its filament structure was built up from scratch using brass wire and filament material taken from an aquarium bulb. As you now know, manual construction of filament assemblies results in fragile, short-lived vacuum tubes.

I used the rubber stopper that comes with the glass tube as a base for the triode. The filament wires were poked through the rubber. The glass drinking tube was removed from the stopper and replaced with a short length of copper tubing. Appropriate fittings were added to allow it to be coupled to my vacuum system. Both the copper tube and the stopper itself were well coated with silicone rubber before being assembled.

The control electrode, the functional equivalent of a grid, was a sheet of copper foil glued around the waist of the bottle. The hamster bottle came with a metal clip with which to attach the bottle to the animal's cage. That same clip, snapped over the foil, provides a convenient means with which to make an electrical connection.

To be honest, my initial results were mixed.

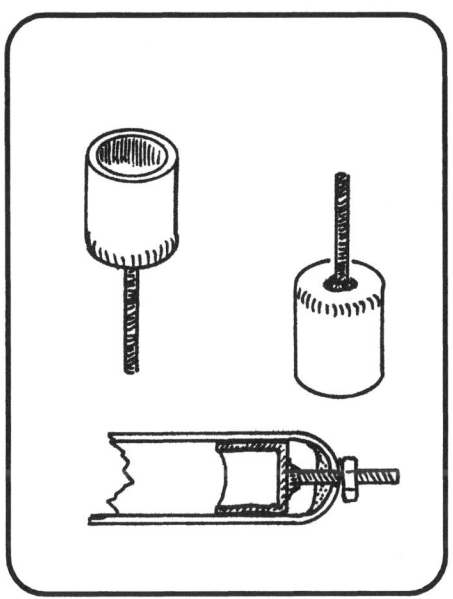

Figure 12.14
Construction and installation of the plate assembly in the Hamster Bottle Triode

Silicone outgasses a great deal more than epoxy, so it was difficult to achieve and maintain the levels of vacuum I was accustom to. The life of the filament was notoriously brief. In fact, the tube was in operation for such a short period of time that I scarcely had opportunity to make any measurements. There was some evidence of amplification, and I even got the tube to oscillate and generate an audio tone in a set of headphones. Shortly thereafter, the filament burned through and the tube went dark.

Page 203

Regrettably, the original HBT is a glued-and-sealed affair. It doesn't lend itself to rebuilding, so the experiment pretty much ended there. If you'd like to tinker with a Hamster Bottle Triode of your own, I would make three recommendations.

First, try to use something other than a rubber stopper as a base. A small, flat plate of glass like a microscope slide might work better. When the time comes, it can be cemented to the mouth of the tube with a minimal bead of silicone. If you need to reopen the tube to effect repairs, the seal can be cut with a razor blade.

A couple of holes will probably have to be drilled through the glass in order to pass the filament wires. Any sealing to be done, either at the filament wire holes, or the mounting hole for the plate, should be done with epoxy.

Finally, don't try to make your own filament. Steal the complete stem assembly out of a low-voltage lightbulb, and use it instead.

Tennis Ball
Vacuum Triode

Pressure:	25 milliTorr
Filament Material:	Tungsten, removed from 12-volt automotive turn/brake lamp
Filament Voltage:	Approx 5 volts
Date:	8/14/00
Note:	Grid voltage applied with respect to negative filament terminal

Grid Voltage Applied	Plate Current (mA) With 50V Plate Voltage	Plate Current (mA) With 45V Plate Voltage	Plate Current (mA) With 27V Plate Voltage	Plate Current (mA) With 18V Plate Voltage	Plate Current (mA) With 9V Plate Voltage
0.00	1.79	1.09	0.11	0.03	0.02
-0.50	1.74	0.95	0.09	0.03	0.02
-1.00	1.64	0.74	0.08	0.03	0.02
-1.50	1.51	0.60	0.07	0.02	0.01
-2.00	1.27	0.46	0.06	0.02	0.01
-2.50	0.99	0.33	0.04	0.02	0.01
-3.00	0.75	0.22	0.04	0.02	0.01
-3.50	0.54	0.18	0.03	0.01	0.01
-4.00	0.39	0.12	0.02	0.01	0.01
-4.50	0.24	0.09	0.01	0.01	0.00
-5.00	0.11	0.03	0.01	0.00	0.00
-5.50	0.03	0.01	0.00	0.00	0.00
-6.00	0.01	0.00	0.00	0.00	0.00
-6.50	0.00	0.00	0.00	0.00	0.00

References

Bucher, Elmer E., *Vacuum Tubes in Wireless Communication*, copyright 1918, Wireless Press, Inc., New York, pp. 62-76, (regenerative amplification systems), pp. 161-167, 173-182, (the Weagant oscillation valve)

Grainger, Maurice J., *Amateur Radio - How and Why of Wireless With Complete Instructions on Operating Receiving Units*, copyright 1922, McClelland & Steward, LTD, Toronto, Canada, p. 65, (comments pertaining to vacuum-tube filament life)

Chapter 13
Semiconductor Basics

For a half century or more, vacuum tubes, in their various incarnations, reigned supreme as the kings of amplification devices. Over those many years, tubes reached a high degree of sophistication, efficiency, and yes, even miniaturization. They were the bedrock on which emerging technologies like computers, aerospace electronics, and television were built. Yet, despite these achievements, and decades of technical enhancement, all vacuum tubes suffered from certain inescapable shortcomings.

As you know, every tube needs a filament or heater of some sort to facilitate the emission of electrons. Tube filaments, like those in common lightbulbs, are susceptible to damage from mechanical shock. Even under ideal conditions, they eventually burn out, rendering the tube useless. As part of a missile guidance system or medical electronics, a failure of this type would have dire consequences.

The energy used to heat the filament, while essential to the operation of the device, represents a loss, because it really contributes nothing to the actual output of the tube. Beyond the energy wasted in heating the filament, tubes require comparatively high plate voltages for operation. For these reasons, tubes don't lend themselves well to battery-powered applications where the power supply is necessarily limited.

Finally, the composition and physical makeup of tubes, including their delicate electrode assemblies, makes them inherently fragile, sensitive to mechanical shock, and prone to breakage.

Long before engineers and scientists had learned to control the flow of electrons through a vacuum, they had already observed curious behavior in the flow of electric currents through certain types of crystalline materials. Unlike obvious conductors like copper, or insulators like glass, these crystals seemed to exhibit some of the characteristics of both. They might conduct an electric current one moment, or resist its flow the next, depending upon the crystal's composition and the manner in which the current was applied. These crystals comprised the first of a new class of materials, now known as *semiconductors.*

As early as the 1890's, J. C. Bose was using semiconducting crystals as diodes with which to detect microwaves. Later, inventors like G. W. Pickard and H. C. Dunwoody applied materials like fused silicon, galena, iron pyrite, and carborundum crystals to radio communication equipment. In 1942, S. Benzer, of Purdue University, discovered the rectification properties of crystal germanium.

In studying these materials, numerous experimenters sought to improve their performance by probing the crystals with additional electrodes, or stimulating them with voltages applied here and there. After all, if the addition of a grid to a thermionic diode resulted in an amplifying device, why couldn't they build a "crystal" triode?

If ham radio lore can be taken at face value, the potential for mineral crystals to amplify was demonstrated on a number of isolated occasions, apparently in several different countries. Whether the observers understood what they were really doing, or appreciated the significance of what they had witnessed is another matter entirely. It can be said with certainty that the conceptual design for a "solid-state" amplifier was patented as early as 1930 (J. E. Lillienfeld, U.S. Patent #1,745,175). In 1938, R. Pohl and R. Hilsch published a paper entitled "Control of Electric Currents With a Three Electrode Crystal and a Model of a Barrier Layer." In it, they describe a potassium bromide crystal, fitted with three electrodes, that was capable of amplifying input currents by a factor of 100.

The poor quality of early semiconducting materials, an inadequate grasp of the physical processes of conduction, and the upheaval connected with World War II retarded the invention of the first practical semiconductor amplifier until 1948. In that year, John Bardeen and Walter Brattain patented their germanium-based crystal triode, or *transistor*, as they dubbed it. The world would never be the same.

Vacuum tubes, which had achieved a high degree of refinement by this time, remained dominant in electronics for several decades more. Slowly, but surely, they were replaced in new designs by transistors and transistor-like devices. In recent years, thermionic tubes have been so completely vanquished by semiconductors that they remain in but a handful of specific applications. In this day and age, few people would recognize a vacuum tube if it hit them in the head.

Semiconducting amplification devices like transistors offer significant advantages over tubes. They require no filament and are capable of operating at very low voltages. This makes them efficient, and ideally suited for use in battery-operated, portable electronics. They're mechanically simple, lightweight, and tolerant of vibration and shock. Compared to tubes, they're small, very small.

L a t t e r - d a y semiconductor technology has shrunk the transistor to the point that millions of them can be fashioned on the head of a pin. Extensive automation, and the economies of mass production have made them unthinkably cheap. Inexpensive, low-cost, extremely miniaturized electronics has made possible virtual miracles in computers, robotics, medical electronics, satellite communications, entertainment electronics and more. In short, 21st-century life as we know it would be impossible without the amplifying capabilities of semiconducting crystals.

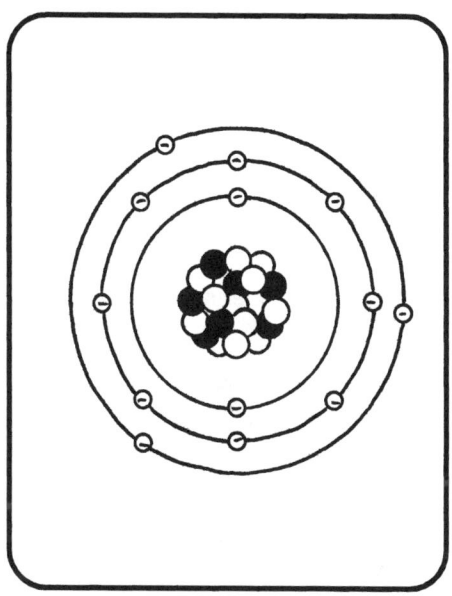

Figure 13.1
A diagrammatic view of an aluminum atom, showing protons (white), neutrons (black), and the orbiting electrons

This book, of course, is about hands-on science. Happily, it *is* possible for the hobbyist or basement engineer to fabricate and experiment with primitive semiconducting materials and devices, including transistors. Before I introduce these experiments, however, the differences between conductors, semiconductors, and nonconductive materials bears some consideration.

Anyone who's spent any time tinkering with electricity can usually guess, with some accuracy, whether or not a given piece of "mystery" material will prove to be a conductor or nonconductor. It's a fair bet that if a sample is metallic, it'll probably conduct electricity. If it's non-metallic, it probably won't. Unfortunately, the obvious external difference between a chunk of copper and a chunk of sulfur, for example, says nothing about the mechanism that prescribes the differences in their conductivity.

Moreover, some materials, like carbon, change their behavior depending upon the form in which they exist. Black, amorphous carbon is a fair conductor of electricity. A diamond, which is carbon in its crystalline form, is about as *non*-conductive as a material can be. How can this be explained?

A Simple Conductor

In Chapter 8, I introduced to you the underlying structure of the atom. (I would urge you to review that chapter, as necessary.) If you recall, all atoms feature a positively charged nucleus composed of protons and neutrons. The number of protons contained in the nucleus determines the identity of the atom. Surrounding the nucleus are "orbiting" clouds of negatively charged particles called electrons.

Under normal conditions, the total number of electrons in orbit precisely matches the number of protons contained in the nucleus. Mathematically speaking, the negative charge on each electron cancels the positive charge on a corresponding proton. Thus, the total charge of the atom as a whole, is zero. Under these conditions, the atom is electrically neutral.

The motion of these electrons is confined to a number of well-defined regions in space called *shells*. The shells are filled according to physical laws that, among other things, determine how many electrons each shell can hold, and the specific order in which shells may be filled. For our purposes, the shells can be thought of as being concentric and nested, like a set of Russian matryoshka dolls.

The electrical field that binds the electrons to their nucleus dissipates rapidly as the distance from the nucleus is increased. Electrons orbiting in shells close to the nucleus tend to remain bound tightly to the atom. Their more distant brothers, orbiting in the outermost shell, are less restrained by these attractive forces. As a result, they're more likely to be influenced by external factors.

As we'll see momentarily, the outer shell and its inhabitants are so important to understanding the physical behavior of atoms, that they've been assigned special names—the *valence shell* and *valence electrons*, respectively.

Figure 13.1 is a stylized depiction of the internal structure of an aluminum atom. Notice that the valence shell contains three electrons. These lone electrons are so loosely bound to the atom that it takes very little additional energy to agitate them. In fact, there's enough latent energy present in a block of aluminum at room temperature to break some of them loose.

As these outer electrons are liberated, they become what are known as *free electrons*, which wander the spaces between adjacent atoms. In a given sample of aluminum, there are trillions of atoms undergoing this process at any given instant.

"Doesn't the loss of an electron upset the electrical balance in the atom?" you ask. Yes. The loss of an electron will cause the atom to assume a net positive charge. However, the condition is temporary. At such an instant as another free electron wanders by, the atom will probably snatch it up, restoring its electrical neutrality.

The process by which valence electrons are shed and sporadically recaptured is essentially random, and more or less continual. The important point to note is that, because of it, there's always a liberal supply of free electrons available.

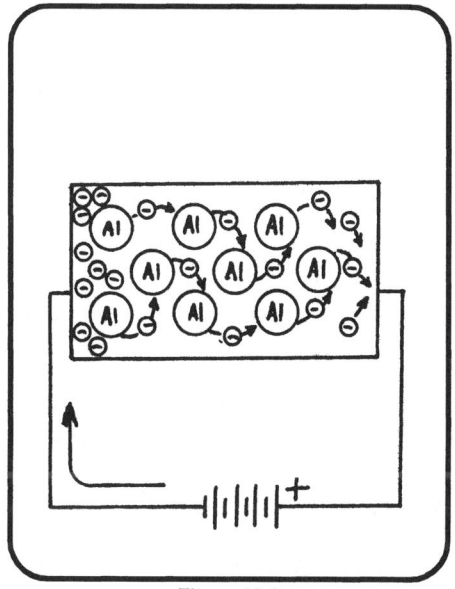

Figure 13.2
The electric field from an external battery sweeps free electrons through a block of aluminum

This brings us to the matter of electrical conduction. If the aluminum sample is fashioned into a wire and then connected to the terminals of a battery, the free electrons can be swept from one end of the wire to the other in a continuous stream.

The negative terminal of the battery deposits swarms of additional electrons into the wire. The aluminum atoms, which are "willing" to release and capture wandering valence electrons, happily pass electrons from one to another. The action is like an old time bucket brigade, wherein each fireman passes a bucket of water to the next man in line. The electrons ripple through the wire, passing from host to host, until they appear at the other end. There, the positive terminal of the battery drains them away. All in all, aluminum is a pretty decent conductor, bested only by silver and copper. See figure 13.2.

While the presence of free electrons serves to explain conduction in a material like aluminum, it does not address why other materials should be nonconductors. One can assume there's a shortage of free electrons in the latter, but this only begs the logical question: "Why do some atoms freely release electrons when others will not?"

Noble Gases

Just over a hundred different types of atoms are known to exist. Atoms tend to be social creatures, predisposed to bind with other atoms to produce *molecules.* The total number of possible combinations is staggering, if not incalculable, and accounts for the incredible diversity in the substances found in this universe.

A grain of table salt, for example, is comprised of simple molecules, each containing a single sodium atom joined to a chlorine atom. Water is the result of the union between an oxygen atom and two hydrogen atoms. Even oxygen gas, which accounts for 21% of Earth's atmosphere, does not exist in the form of isolated atoms. Rather, it's made up of oxygen molecules, each of which is composed of two oxygen atoms, bonded together. As a rule, atoms enjoy company.

Interestingly, six elements are exceptions to this rule. Collectively referred to as the *noble gases,* they include helium, neon, argon, krypton, xenon, and radon. Like aloof aristocrats, who would not be seen in the company of commoners, so too, these elements refuse to interact with other atoms. As a result, for all intents and purposes, they are chemically inert.

If we examine the shell configurations for each of the noble gases, a curious pattern emerges. In each case, the valence shell is either full, or it contains eight electrons (sometimes referred to as the *octet rule*). Review the table entitled "Electron Configurations for Noble

Gasses," which can be found at the end of this chapter. Suffice it to say, these configurations are highly stable, and therefore, very desirable.

As to their chemical inertness, noble gas atoms are, by definition, both electrically neutral and electronically stable. They will not engage other atoms, because there's nothing to be gained by a transaction with another.

Simple Compounds

The table entitled "Electron Configurations for Sodium and Chlorine" shows the electronic structure for several select atoms. Sodium features 11 protons in its nucleus, with 11 electrons in orbit around it. Its valence shell contains only 1 electron. As such, the valence shell is neither full, nor does it meet the criterion for the octet rule. Consequently, though the atom is electrically neutral, its electron configuration is not optimally stable.

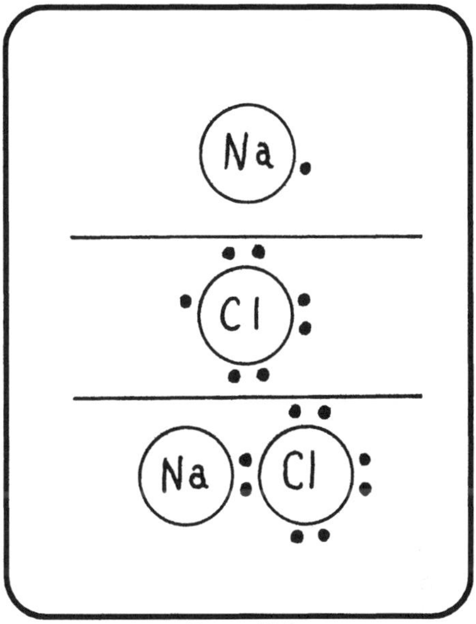

Chlorine, with 17 protons and electrons, has 7 electrons in its valence shell. It, too, is electrically neutral, but like the sodium atom, its valence shell is neither full, nor an octet.

Figure 13.3
Lewis diagrams for sodium, chlorine, and a sodium chloride (table salt) molecule

Both atoms would prefer to have the highly stable electron configurations like those associated with the noble gases. Through a curious process of *lending* and *borrowing*, both of their configuration wishes are realized when the two are brought together.

To be concise, chlorine captures sodium's valence electron, making it one of its own. With the added electron, chlorine's valence shell satisfies the octet rule. No wonder, as its electronic configuration now looks like the configuration for the noble gas argon!

One would think sodium sorry to have lost its valence electron, but the deal is a win-win situation. Without that electron, sodium's electron configuration now looks just like the configuration for the noble gas neon!

Let's not forget that with the loss of its valence electron, sodium's electrical balance is upset. Without it, there are no longer sufficient electrons to offset its protons, so the atom assumes a net positive charge of (+1). Conversely, when chlorine takes ownership of that electron, its balance is also upset, but in the opposite direction. The additional electron changes chlorine's net charge to (-1).

Opposite charges attract. So, by virtue of their opposite electrical charges, the sodium and chlorine atoms attract each other and bind into a sodium chloride molecule. The end result is a highly stable molecule that tastes great when sprinkled on a plate of french fries.

To summarize, we can say that some electron configurations are more stable than others. Atoms will lend and borrow electrons, to varying degrees, if such action results in a more stable electron configuration for each. This lending and borrowing forms the basis for the bonding between two or more atoms, and allows for the formation of a wide variety of molecules.

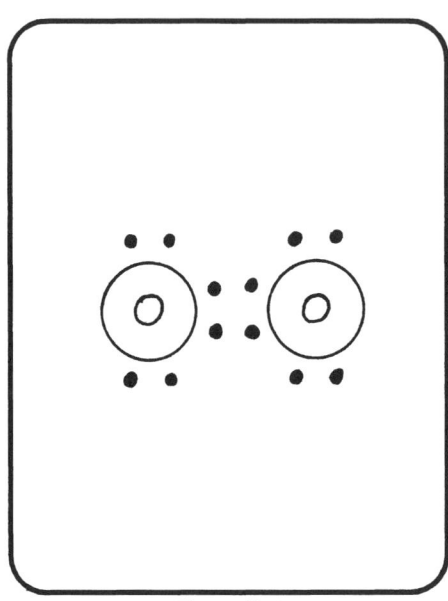

Figure 13.4
Covalent bonds are found in molecular oxygen

Since the valence electrons are the only electrons involved in chemical bonding, drawing *all* of the orbits present in a given atom is really overkill. Chemist Gilbert Lewis proposed a more efficient diagram scheme, wherein he drew no more than a vague nucleus surrounded by a series of dots, one for each valence electron. Sodium, then, becomes a nucleus with one orbiting dot. Chlorine is drawn as a nucleus with seven dots. The Lewis diagram for sodium chloride (table salt) implies chlorine's theft of sodium's valence electron. See figure 13.3.

Covalent Bonds And Crystalline Carbon

The marriage of sodium to chlorine is a useful example of molecular bonding. In essence, it's really the product of an electron transfer or theft, facilitated by each atom's obsessive desire for a stable electron configuration. Chemists offer the more polite phrase, ionic bonding, to describe this type of relationship.

Most chemical bonds are far less extreme. Many atoms, because of their existing configurations, find that they can satisfy their desires for electronic "nobility" through more cooperative means.

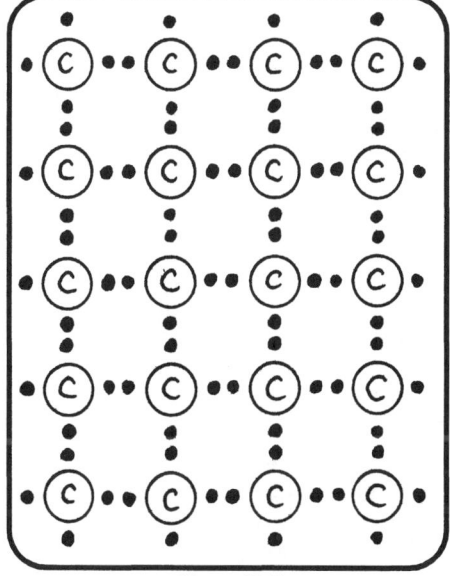

Figure 13.5
Lewis diagram of crystalline carbon

Oxygen gas is a good example of cooperative bonding. Oxygen atoms, as individuals, each have six valence electrons. Eight electrons would satisfy the octet rule, giving them a desirable neon-like electron configuration. Oxygen atoms, however, are not inclined to steal from one another. Instead, they bond together in couples, each *sharing* two electrons with the other. See figure 13.4. Cooperative bonds of this type are called *covalent bonds*.

Carbon is an element that ranks among the most abundant in the universe. The pure, black, amorphous variety is composed of an unstructured mass of atoms. It's electrically conductive, and has use in applications ranging from arc lamp electrodes and motor brushes to dry cells. If you recall, it was also a key material from which to build components for the instruments described in Chapters 5 and 6. Like aluminum, amorphous carbon owes its conductivity to its valence electrons, which are easily stripped loose to become free electrons.

Under the influence of intense heat and pressure, a jumbled batch of carbon atoms can be forced into an orderly, repeating pattern. The resulting structure is called a *crystal lattice*, and this particular type of crystal is known to all as the *diamond*. Strangely, while amorphous carbon is a relatively good conductor, diamonds are absolutely not! What happens to carbon in its crystalline state that prevents the release of free electrons?

Figure 13.5 is a Lewis diagram depicting the atoms in a sample of crystalline carbon. Individual carbon atoms have but four valence electrons. In the diamond crystal, however, every atom is able to satisfy the octet rule through covalent bonding (cooperative sharing) with each of four neighbors.

The fact that every available electron is shared and used somewhere to form a link between adjacent atoms means that, in the diamond, no electrons are available to be set free.

Bear in mind that the Lewis diagram is two-dimensional, because it's printed on a flat sheet of paper. Real crystals, like diamonds, have height, width, and depth. They are three dimensional creatures. The repeating pattern of atoms, and the bonds that link them, likewise extend in three dimensions. The signature tetrahedral shape associated with diamonds is the direct product of the geometric relationship between the atoms in the crystal lattice.

Incidentally, carbon is but one member in a family of similar elements, each of which features four valence electrons. This family includes such elements as germanium and silicon. Like carbon, each has some capacity to conduct electricity when in its amorphous state. The extensive covalent bonding in crystalline samples ties up the valence electrons, leaving them unavailable to become free electrons. Hence, they become poor conductors. Germanium and silicon are of special interest to us, as they are the materials from which most contemporary semiconductors are fabricated.

N-Type Impurities and Electrons

Germanium is a grayish-white metalloid, discovered in 1886 by C. Winkler. In nature it's a fairly rare element, but it can be found and extracted in trace amounts from copper, silver, and zinc ores. Like carbon, germanium has four valence electrons. When crystallized, its atoms bond and arrange themselves into patterns very reminiscent of the diamond. Figure 13.6 shows the bonds in a crystal of pure germanium.

A quick look confirms that all the valence electrons are tied up in bonds linking one atom to the next. There are no electrons that can convert easily to the free state, so crystalline germanium, like the carbon diamond, is a poor conductor.

Impurities, added to the germanium during the crystallization process, can introduce defects in the lattice that yield striking results. Figure 13.7 depicts a section of another germanium crystal, this one having been contaminated with a tiny amount of the element arsenic. During the crystallization process, the arsenic atom shown attempted to bond with, and integrate itself, into the crystal structure. For the most part, it was successful, though its valence electrons have caused a slight problem. You see, arsenic is a *pentavalent* atom, meaning that it's got five valence electrons, not four.

In order to fit into the crystal lattice, arsenic utilizes four of the five to bond with neighboring atoms, leaving the

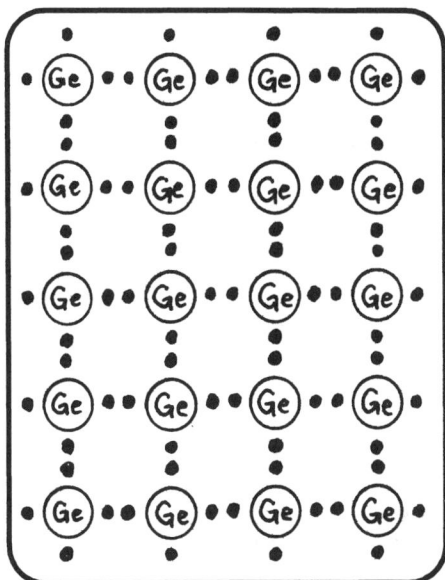

Figure 13.6
A crystal composed of pure germanium

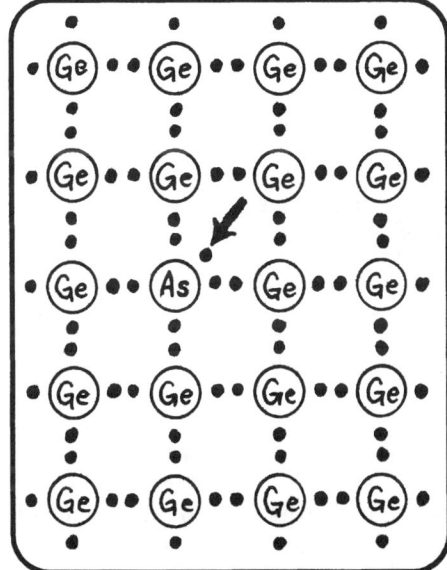

Figure 13.7
A germanium crystal doped with arsenic contains free electrons, making it a N-type material

Page 217

remaining electron un-bonded and uncommitted. With no obligations elsewhere, that "extra" electron is prone to break loose and wander through the crystal as a free electron.

If the tainted crystal is somehow connected to the terminals of a battery, the free electrons will move through the crystal in the same bucket-brigade-like manner that they do in a piece of wire. Magically, the contaminated germanium allows for the flow of electrons, where the pure crystal does not.

An electric current flowing through any material is expressed as the movement of electric charges. In the case of the impure germanium, the flow is implemented in the movement of free electrons. Stated another way, the *charge carriers* in the impure crystal described above are negatively charged. Crystals of this type are called *N-Type* germanium.

The purposeful contamination of germanium with other atoms is called *doping*. Doping germanium with pentavalent atoms causes defects in the crystal that pepper it with free electrons. Arsenic was described in the example, though antimony can be used to the same effect. Both are collectively referred to as *donor impurities*, because their introduction into the crystal lattice results in a "donation" of free electrons, and the creation of an N-Type material.

P-Type Impurities and Holes

The fact that doping changes the electrical characteristics of pure, crystalline germanium suggests other possibilities. For example, indium is a *trivalent* atom, meaning that it has only 3 valence electrons. This, of course, is one *less* than germanium. What happens if a pure germanium crystal is doped with a small quantity of indium?

Figure 13.8 depicts the results. Despite a willingness to share what it's got, indium simply doesn't have sufficient valence electrons to bond with all of its neighbors. It can establish three bonds easily enough, but the lack of sufficient valence electrons means that, for the moment, the fourth bond cannot be realized. From indium's standpoint, the situation is unacceptable. The only way it can come up with the required fourth electron is to steal it from elsewhere. As we can see, a nearby germanium atom becomes the victim of a valence electron theft. Using the stolen electron as its own, indium forges the fourth bond with its neighbors, and secures its place in the crystal.

This is well and good for the indium atom, but what of the victimized germanium atom? It's now missing an electron, and the bond that electron previously implemented is now corrupted. The resulting defect, the region in space where a bond should exist but does not, is called a *hole*. Holes have some interesting properties.

The first thing to consider is that a germanium atom, before being burglarized, is an electrically neutral atom. The theft of an electron upsets this balance, causing the atom to assume a (+1) charge. The appearance of a positive charge is therefore associated with the creation of a hole.

There's no reason that a hole can't be backfilled. If the victimized germanium atom can, itself, steal an electron from another germanium atom, its hole will have been filled, and its charge neutralized. The process, of course, results in the creation of a new hole, associated with the new victim, elsewhere. Notice that this provides a mechanism for the apparent migration or movement of positively charged holes through the crystal. See figure 13.9.

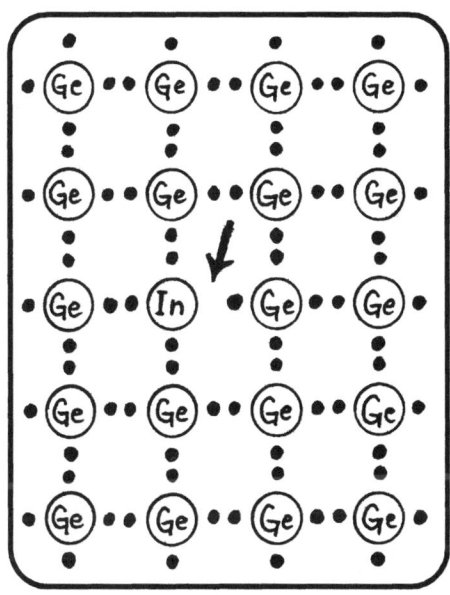

Figure 13.8
A germanium crystal doped with indium features holes, making it a P-type material

There's no need to get wrapped around the axle trying to grasp the intricacies of hole movement. We know holes have a positive charge, and that they can effectively move through the crystal. For our purposes, it's easiest to think of them as virtual particles, similar to electrons, but opposite in polarity.

If a germanium crystal, doped with indium, is attached to the terminals of a battery, a current will flow. The current, however, will not be expressed as the movement of free electrons, because there are none. Instead, positively charged holes move through the crystal in a flow. In this instance, the doped crystal is considered to be *P-Type* material.

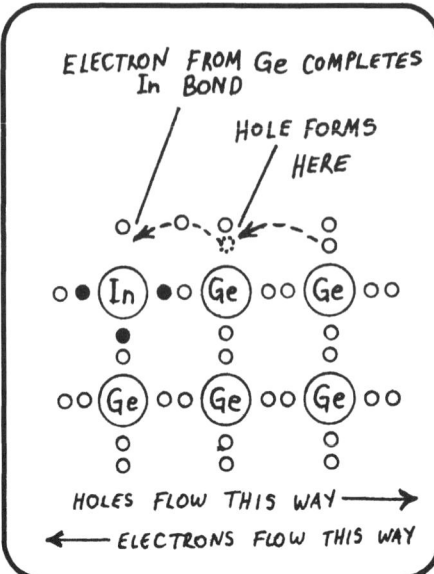

Figure 13.9
The relative movement of electrons and holes
through a germanium crystal

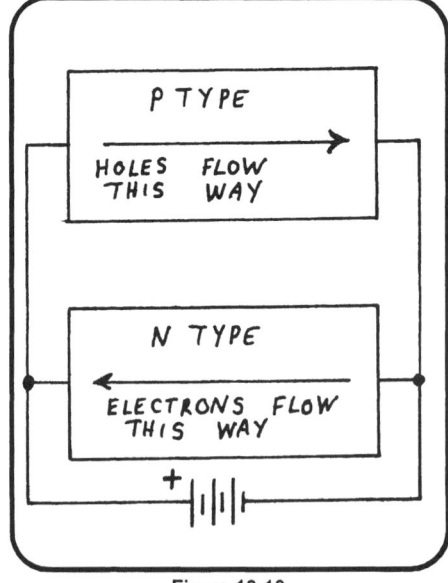

Figure 13.10
Direction of current flow depends upon the type
of charge carrier

If the current flow in an N-Type crystal is compared to the flow of current in a P-Type crystal, an important difference will be observed. The charge carriers in each instance flow in opposite directions. This stands to reason. Electrons, being negatively charged, will tend to flow toward the positive terminal of the battery. Holes, being positively charged, will migrate toward the negative terminal of the battery. See figure 13.10.

Doping germanium with trivalent atoms causes defects in the crystal lattice that riddle it with positively charged holes. Indium was offered as a typical impurity, though gallium can be used to achieve the same result. Both are collectively referred to as *acceptor impurities,* because their introduction into the crystal lattice results in holes, which greedily "accept" electrons in an effort to reestablish their electrical neutrality. The doping of a pure crystal with acceptor impurities yields P-Type material.

A Simple Semiconducting Junction

The preceding pages in this chapter have been filled with discussions of atoms, covalent bonds, lattice defects, holes, valence electrons and more. For those patient enough to have waded with me through these technicalities, the payoff is about to come. Here is where details converge to form a basic, but useful, semiconducting structure called a *junction diode*.

Conceptually, a junction diode is composed of two pieces of crystalline material, one doped to be N-type, and the other doped to be P-type. We can assume for the moment that the material is germanium, and that the two pieces have been doped with arsenic and indium, respectively. The two pieces are brought into contact, forming the P-N "sandwich" shown in figure 13.11. How does this structure respond to an externally applied electric current? That depends upon *how* the voltage is applied.

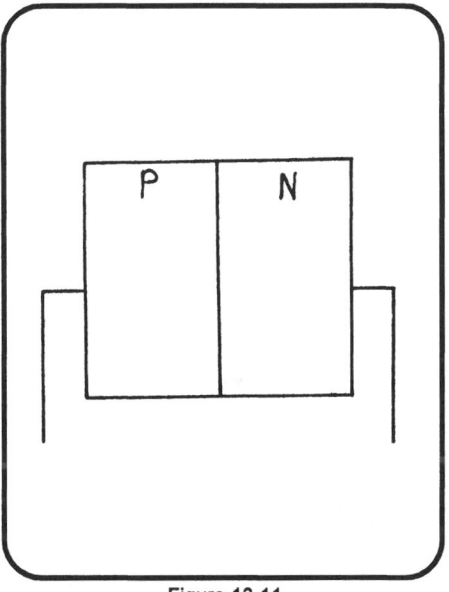

Figure 13.11
A simple P-N semiconductor junction

Figure 13.12 shows the P-N junction connected in series with a battery and a small lamp. Observe that the positive terminal of the battery is attached to the "P" side of the junction and that the negative terminal of the battery, having been routed through the lamp, is connected to the "N" side of the junction. Will the bulb light?

Electrons, leaving the negative terminal of the battery, pass through the bulb and flow into the "N" side of the junction. The free electrons, already present in N-type material, are driven by mutual repulsion toward the P-N boundary. There, they eventually cross over to the P-type material on the other side. Despite the fact that electrons are escaping from the "N" to the "P" side of the junction, the "N" side

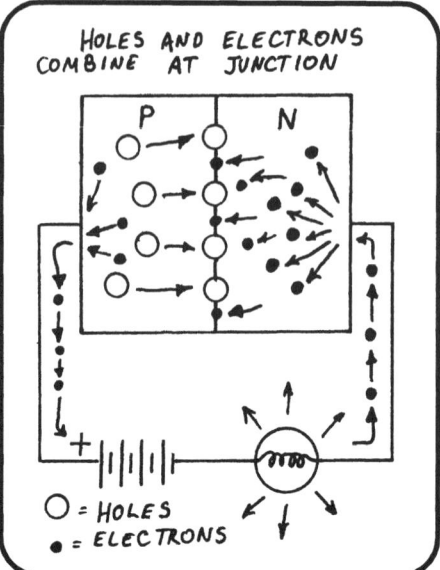

Figure 13.12
Electric current flows freely through a
forward- biased P-N junction

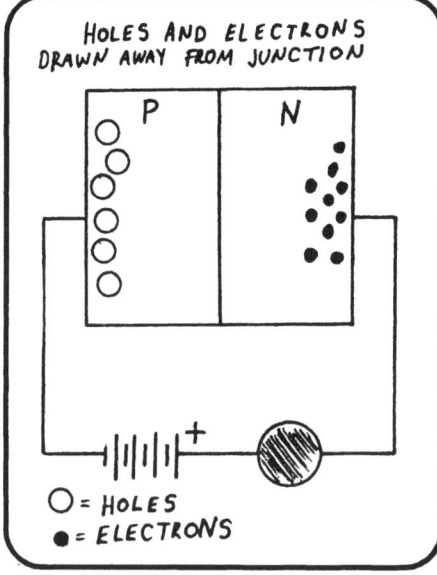

Figure 13.13
No electric current flows through a
reverse-biased P-N junction

never runs out, because the wire from the battery supplies an endless stream of new ones.

Electrons that have crossed into the "P" side of the junction are strongly attracted to the holes they find there. In fact, both the electrons and holes effectively vanish, because they recombine to form neutral atoms.

The P-type material might be expected to eventually run out of holes, were it not for the fact that it's connected to the positive side of the battery. The positive terminal of the battery attracts electrons out of the "P" material. Every electron drawn out of the P-type crystal by the battery effectively creates a new hole. Newly generated holes migrate across the P-type material toward the P-N border, where they become available to combine with electrons arriving from the "N" side of the P-N junction.

Internally, a stream of electrons moves toward the P-N boundary from the negative side, while a stream of holes drifts toward the boundary from the positive side. As long as the external battery supplies a potential, the movement of charge carriers in the P-N sandwich is continuous.

If we ignore the processes internal to the diode, we can see that electrons flow out of the battery, through the bulb, and into the "N" side of the diode. They appear at the "P" side, and return to the battery. The circuit is complete, and flow of electrons is continuous. The lamp will, indeed, light.

When a P-N junction is connected to an external power source in such a manner that charge carriers are driven toward the junction boundary to combine, the junction is said to be *forward-biased*. A P-N junction biased in the forward direction will support the flow of an electric current.

What happens if the polarity of the power supply is reversed, so that the positive terminal of the battery is connected to the "N" side of the junction, and the negative terminal of the battery connects to the "P" side?

In this case, the junction is said to be *reverse-biased*. The positive terminal of the battery draws free electrons present in the N-type material away from the P-N boundary. The negative terminal of the battery draws holes away from the boundary in similar fashion. With the holes and electrons segregated and unavailable to move and traverse the P-N boundary, the junction becomes, for all intents and purposes, an insulator. No available charge carriers means that no current can flow. So, a P-N junction, biased in the reverse direction, will not support the flow of an electric current. The lamp depicted in the circuit in figure 13.13 will remain dark.

Note that the P-N junction diode, like the thermionic diode described a few chapters back, allows the flow of current in one direction, but not the other. Unlike the vacuum tube version, it performs this task without the need for a filament.

Conclusion

The thermionic diode is not an instrument of amplification. The thermionic *triode*, is. The triode, with its control grid, represents a more advanced implementation of concepts that make the thermionic diode work.

Likewise, the P-N junction just described is not, in and of itself, an amplifying device. It functions simply as a one-way valve for the passage of electric current. More complex arrangements of P-N junctions, however, yield transistors, which can amplify.

Electron Configurations for Noble Gasses
(Valence Electrons in Bold)

Element & Atomic Number	Shell (1)	Shell (2)	Shell (3)	Shell (4)	Shell (5)	Shell (6)
Helium (2)	**2**					
Neon (10)	2	**8**				
Argon (18)	2	8	**8**			
Krypton (36)	2	8	18	**8**		
Xenon (54)	2	8	18	18	**8**	
Radon (86)	2	8	18	32	18	**8**

Electron Configurations for Sodium and Chlorine
(Valence Electrons in Bold)

Element & Atomic Number	Shell (1)	Shell (2)	Shell (3)	Shell (4)	Shell (5)	Shell (6)
Neon (10)	2	**8**				
Sodium (11)	2	8	**1**			
Chlorine (17)	2	8	**7**			
Argon (18)	2	8	**8**			

Electron Configurations for Carbon-Like Elements
(Valence Electrons in Bold)

Element & Atomic Number	Shell (1)	Shell (2)	Shell (3)	Shell (4)	Shell (5)	Shell (6)
Carbon (6)	2	**4**				
Silicon (14)	2	8	**4**			
Germanium (32)	2	8	18	**4**		

References

Headquarters, Department of the Army, *Basic Theory and Application of Transistors*, copyright 1963, Dover Publications, Inc., New York, pp. 12-39, (semiconductor physics)

Coblenz, Abraham, and Owens, Harry L., *Transistors: Theory and Applications*, copyright 1955, McGraw-Hill Book Company, Inc., New York, pp. 16-19, (holes), pp. 45-58, (semiconductor physics), pp. 58-72, (point-contact transistors), pp. 73-89, (junction transistors)

Dewitt, David and Rossoff, Arthur L., *Transistor Electronics*, copyright 1957, McGraw-Hill Book Company, New York, pp. 337-338, (point-contact transistor details)

Eckert, Michael, and Schubert, Helmut, *Crystals, Electrons, Transistors*, copyright 1990, American Institute of Physics, New York, pp. 106-109, (crystal triode experiments with potassium bromide)

Hunter, Lloyd P., Editor, *Handbook of Semiconductor Electronics*, copyright 1956, McGraw Hill Book Company, Inc., New York, pp. 1.1-1.3, (point-contact transistor characteristics), pp. 1.3-1.10, (junction and field effect transistor characteristics), pp. 4.17-4.20, 4.27, (details of point-contact transistor construction and operation), pp. 5.3-5.8 (photoconductivity in point-contact devices), pp. 8.7-8.8, (metal-to-semiconductor contacts, and forming details), p. 9.7, (encapsulation and effects of moisture on junctions)

Kiver, Milton S., *Transistors*, copyright 1962, McGraw-Hill Book Company, Inc., New York, pp. 4-30, (semiconductor physics), pp. 30-46, (junction behavior), pp. 46-50, (point-contact transistor detail)

Krugman, Leonard, *Fundamentals of Transistors*, copyright 1954, John F. Rider Publisher, Inc., New York, pp 1-8, (basic semiconductor physics), pp. 9-11, (operation of point-contact transistors), pp. 10-15, (operation of junction transistors)

Lange, Norbert Adolph, PH. D., *Handbook of Chemistry*, copyright 1956, Handbook Publishers, Inc., Sandusky, Ohio, pp. 67-68, (physical properties of germanium), pp 83-84, (physical properties of silicon)

Nathan, Harold D., Ph.D., *Cliffs Quick Review Chemistry*, copyright 1993, Cliffs Notes Incorporated, Lincoln, Nebraska, ISBN 0-8220-5318-7, pp. 31-37, (electron orbitals and valence electrons), pp.39-43, (lewis diagrams) pp. 44-45, (the nature of ionic bonds)

Pollack, Harvey, *Transistor Theory & Circuits Made Simple*, copyright 1958, American Electronics Co., Mineola, L.I., New York, pp. 15-23, (operation of junction transistors), pp. 1-12, (semiconductor physics), pp. 12-13, (invention of germanium diode and transistor), pp. 23-28 (operation of point-contact transistors)

Schure, A., *Basic Transistors*, copyright 1961, Hayden Book Company, Rochelle Park, New Jersey, pp. 4-18, (semiconductor physics)

Spenke, Eberhard, *Electronic Semiconductors*, copyright 1958, McGraw-Hill Book Company, New York, pp. 3-54, (semiconductor physics), p. 73, (cuprous oxide semiconductors), pp. 74-81, (metal-to-semiconductor contact)

Torrey, Henry C. and Whitmer, Charles A., *Crystal Rectifiers*, copyright 1964, Boston Technical Publishers, Inc., Lexington, Massachusetts, (details of construction and behavior of point-contact crystal rectifiers)

Towers, T.D., and Libes, S., *Semiconductor Circuit Elements*, copyright 1977, Hayden Book Company, Rochelle Park, New Jersey, pp. 1-13, (semiconductor device fundamentals)

Wailes, Raymond B., Editor, *Manual of Formulas*, copyright 1932, Popular Science Publishing Co., New York, reprint copyright?, Lindsay Publications, Bradley, IL, ISBN 1-55918-036-6, p. 132, (synthetic radio crystals)

"Invention of the Solid State Amplifier," Virgil E. Bottom, *Physics Today*, February, 1964, pp. 24-26, (the inventions of J. Lillienfeld)

"More on the Solid-State Amplifier and Dr. Lilienfeld," *Physics Today*, May, 1964, pp. 60-62

http://www.tuc.nrao.edu/~demerson/bose/bose.html
Emerson, D.T., "The Work Of Jagadis Chandra Bose: 100 Years of MM-Wave Research," copyright ?, National Radio Astronomy Observatory, last revised February, 1998

Chapter 14
The Plumber's
Point-Contact Transistor

The discussion of the germanium diode, as it appears in the last chapter, is really more conceptual than practical. In addition to the tangible difficulties associated with obtaining, purifying, doping, and crystalizing germanium, there is the non-trivial matter of creating the junction itself. Simply pressing two doped crystals together will not achieve the desired results. The junction diode was not a realizable device until suitable material processing and fabrication technologies emerged.

The Point-Contact Diode

Happily, there are countless combinations of materials from which primitive, diode-like junctions can be created. At the dawn of the age of radio, at the beginning of the last century, these tended to be naturally occurring mineral substances like galena or iron pyrite. While there are different ways to combine such materials to produce crude P-N junctions, most radio detectors were composed of a small sample of the crystalline mineral, probed by a thin, sharp, springy wire, whimsically referred to as the *cat's whisker*. The body of the crystal provided one terminal for the resulting diode. The cat's whisker provided the other.

Ideally, a P-N junction allows the free flow of electricity when forward-biased, and completely blocks that flow when reversed. When they were the state-of-the-art, natural mineral diodes offered acceptable performance in simple radio circuitry. Virtually all of them,

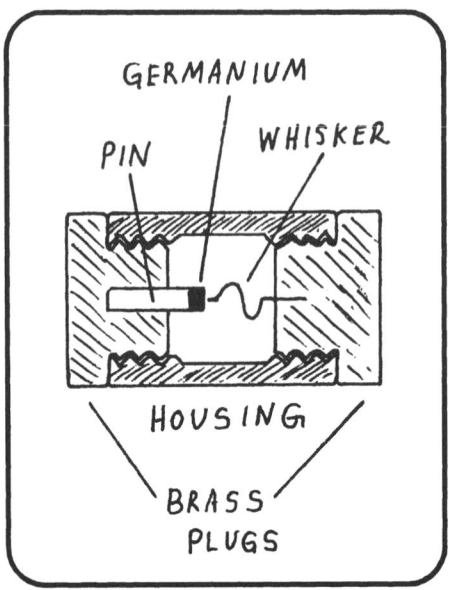

Figure 14.1
Internal construction of a vintage germanium
diode

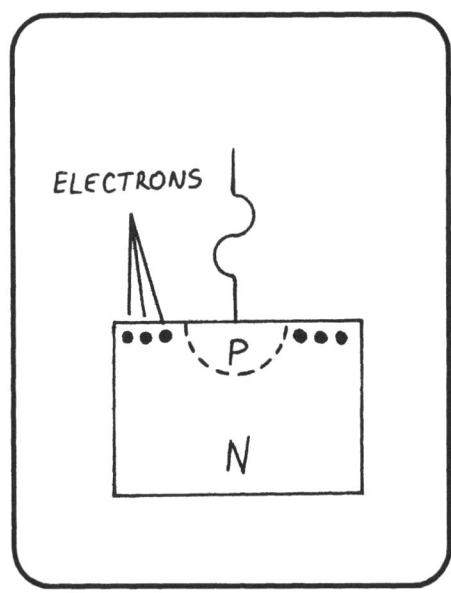

Figure 14.2
The point contact-diode

however, were less than ideal from an electrical standpoint. Many were poor conductors when forward-biased, and most tended to "leak" when reverse-biased.

When purified doped germanium crystals became available, they were ultimately probed with whiskers in the traditional style, and found to make exceptional diodes.

Figure 14.1 is a drawing of an early style of germanium crystal diode and its internals. The diode is composed of a tiny slab of N-type germanium mounted to the end of a terminal post. A tiny "S"-shaped whisker, connected to a second terminal post, probes the surface of the germanium slab. The wire from which the whisker is made is springy. The "S"-shape ensures that the point of the whisker is applied to the face of the crystal with a uniform force. The constituent parts are housed in a ceramic tube that insulates the terminals from each other, and protects the diode's delicate internals.

If you've been paying attention, you should have noticed that I've made no mention of P-type material. How can one create a P-N junction using N-type material alone?!

The answer lies in figure 14.2. A good number of the free electrons in the N-type material diffuse to the surface of the material where they form a virtual skin. Metals, like those used in the fabrication of cat's whiskers, are excellent conductors of electricity. When the whisker is brought into contact with the N-type germanium, some of the crystal's free electrons, near the immediate point of contact, wander into the whisker. The loss of electrons from the vicinity of the whisker upsets the local charge balance, causing the formation of a small, dome-shaped P-type region in the crystal.

Because the P-N junction is implemented as a cat's whisker that makes contact with the face of the crystal at one, tiny point, the configuration is called a *point-contact diode*.

The Crystal Triode

Figure 14.3 illustrates the conceptual design for a point-contact triode, similar to the first transistor developed by Bardeen and Brattain. Like its lesser cousin, the crystal diode, the triode begins with a slab of N-type germanium, which we'll hereafter refer to as the *base*.

The base is probed with not one, but two cat's whiskers. One whisker, labeled *emitter*, is connected to a small battery and a switch. The other whisker, called the *collector*, is wired to a battery of its own, and a current indicator like an ammeter.

If you study the manner in which the collector battery is wired, you'll discover that the diode junction formed by the base and collector is, in fact, reverse-biased. There are no charge carriers available, so virtually no current will flow, with the exception of the tiniest bit of leakage. For all practical purposes, the meter will read zero.

The diode formed by the emitter and base, on the other hand, has been wired in the forward-biased direction. When the switch is closed, electrons are sucked out of the emitter by the battery. The aggressive removal of electrons, and thus the removal of negative charges in this region, results in the creation of holes. In fact, the process is sometimes cleverly referred to as *hole injection.*

These new holes are attracted to the negative field surrounding the collector and begin to travel in that direction. Because they're wandering through N-type base material, which is rife with free electrons, a good percentage of them will eventually encounter an electron and combine with it. Those holes are essentially filled, and cease to be of interest to us.

However, if the emitter (where the holes are injected) and the collector (to which the holes are drawn) are spaced closely enough, significant numbers of holes will reach the collector before being neutralized.

The surviving holes form a positive space charge which attracts and lures additional electrons from the bulk of the N-type base material. With the sudden availability of electron charge carriers in the vicinity of the collector, the ammeter needle will rise, because a relatively large current can now flow through the previously unpassable collector/base junction.

In concentrating on the movement of minuscule charges, it's easy to overlook the overall significance of the process I've just described. Let me rephrase it in the context of the earlier chapters in this book.

A while back, we defined an amplifier, in its most generic form, as a device that allows one source of power to control another. We cited the telegraph relay, as an example. In that case, a small signal applied to the coil of the relay caused it to activate a much stronger local supply. Likewise, our thermionic triodes were constructed so that a delicate voltage applied to the tube's grid was capable of controlling the power source attached to the tube's plate.

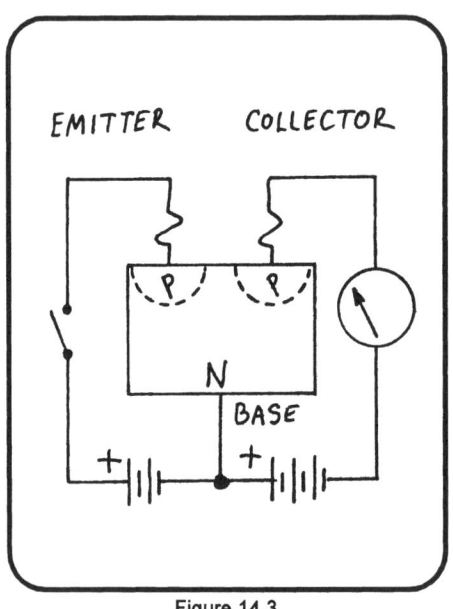

Figure 14.3
The point-contact transistor

In the case of the crystal triode, the battery and meter in the collector branch of the circuit are reverse-biased and normally off. That is to say, there's no significant flow of electric current through that branch of the circuit. When a small current is applied to the emitter branch, the collector branch suddenly comes alive. One power source, the emitter battery, is able to control a second source, wired through the collector. The crystal triode, or point-contact transistor, if you will, is an instrument of amplification.

The nice thing about transistors is that they're not limited to on-off applications like primitive relays. They're equally effective in regenerating and strengthening variable, analog-type signals in the manner of vacuum tubes. In terms of gain, the output of a typical point-contact transistor can double or even triple its input.

The transistor, in its point-contact form, made its debut in the late 1940's. In researching the subject, I collected and read a fair assortment of books from the 1950's, and discovered considerable disagreement between the theories proposed to explain the internal operation of these devices. To complicate the matter, the reign of the point-contact transistor was, by all accounts, brief, and by the early 1960's, they had already been rendered obsolete and extinct by the introduction of superior junction transistors. References to the former vanished from the pages of textbooks from that time forward.

I have no doubt that modern physics has developed the tools and theoretical underpinnings to describe, conclusively, the processes internal to point-contact devices. What's not clear is who, if anyone, has ever revisited the matter. I mention this simply to warn you that you may find my explanation at odds with another period source, a contemporary source, or even with reality. How's that for a caveat?

Homemade Point-Contact Transistors

Almost as soon as the invention of the point-contact transistor was announced, clever experimenters, of the type likely to read this sort of book, sought to fabricate their own. In fact, I was able to locate no less than three magazine articles on the subject. Surprisingly, the process is simple. I was able to glean the essential details, both from these articles and other textbook material, and build several transistors myself.

To build a successful transistor yourself, several issues must be addressed. First, is the obvious matter of securing a sample of N-type germanium. All of the authors I read based their designs on the tiny slab of germanium found in the vintage 1N34 crystal diode. Back then, the 1N34 was manufactured in the type of ceramic package described earlier, in figure 14.1. It was a fairly simple matter to dismantle the housing, or if need be, crush the ceramic tube. If care was exercised, a squat, brass pin, on which the germanium slab or die was soldered, could be safely extracted.

The next concern is the matter of cat's whisker electrode material. Most of the texts I've read call for the use of phosphor bronze wire. One experimenter built his devices using fine tungsten, which he had salvaged from the filaments of burned-out 35W4 vacuum tubes. In any event, the electrode material must be of an extremely fine gauge, or must be capable of being honed to a sharp point. It must also have some elasticity or spring-like qualities, so that when the points are placed upon the surface of the germanium, they'll tend to remain there under a delicate but uniform pressure.

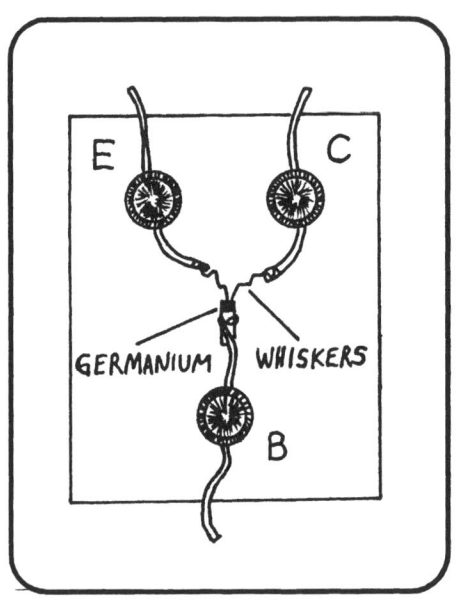

Figure 14.4
Atkins' homebrew transistor

The third concern, I'll describe as a matter of geometry. The operation of the point-contact transistor is dependent upon the successful migration of holes from the emitter to the collector. There are plenty of free electrons in the N-type germanium to intercept them, so the only way to ensure that at least some holes make their way to the collector is to place the emitter and collector whiskers in close proximity. All of the references I found described emitter-to-collector distances ranging from 0.0005 to 0.005 inch, with 0.002 being the most common value.

A successful transistor can be fabricated in any number of ways, provided the elements above are combined in an appropriate manner. This, of course, is largely a question of mechanics. However it may be implemented, the body of the transistor must somehow contain the germanium slab, provide support for the emitter and collector cat's whiskers, and keep them applied under light pressure to the face of the germanium crystal. All the while, the whiskers must be maintained at their proper spacing. The authors of each of the three construction articles I found implemented their transistor housings in different ways.

C. E. Atkins (see references) took a rather straightforward, "breadboard" approach, the essence of which can be found in figure 14.4. In his design, a block of wood features three brackets with binding posts. One binding post supports the crystal itself, while each of the other two supports a cat's whisker. His scheme is simple, easily handled and adjusted, and apparently works. The drawback I see has to do with an obvious lack of mechanical rigidity. The whiskers are easily dislodged from their delicate adjustment by the slightest vibration or shock. As well, Atkins's crystal is completely unprotected from dust and other airborne contaminants.

Apparently, experimenter/author Rufus Turner viewed the transistor, quite literally, as a diode with two whiskers. Turner's approach to transistor construction involved the modification of an old ceramic 1N34 to accommodate the additional whisker. The implementation, as he described it, is quite clever,

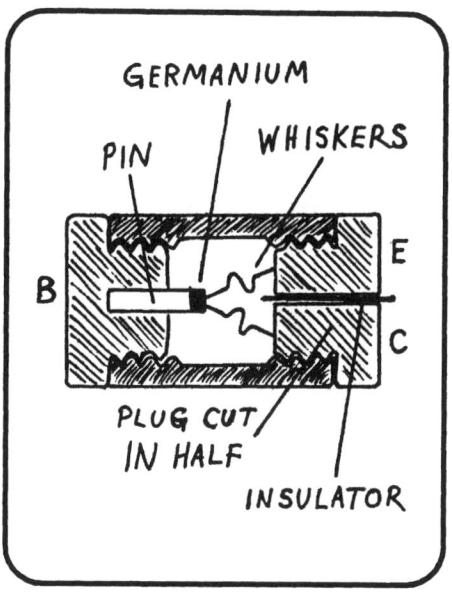

Figure 14.5
Turner's homebrew transistor

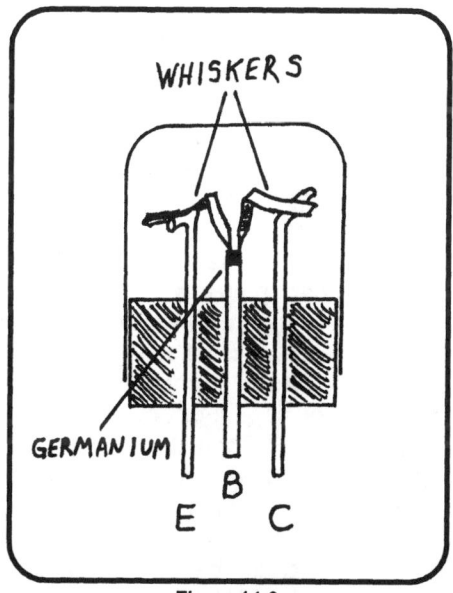

Figure 14.6
Helsdon's homebrew transistor

Page 235

but is bound to be problematic for the contemporary experimenter. For mechanical reasons, it requires not one but *two* of the vintage-type ceramic 1N34's. Vintage 1N34's may still be found, but I'd hardly consider them plentiful. A design requiring the parts from two of them is obviously disadvantageous. As well, I've puzzled over his ability to dismantle them in a non-destructive fashion. Despite my best efforts, I have yet to succeed in opening a ceramic 1N34 without fracturing it. Figure 14.5 summarizes Turner's design.

To me, P. B. Helsdon's design is the most practical, and is most likely to result in a useable device. Helsdon's design begins with a tiny cylinder of resin or similar plastic insulating material, which serves as the foundation of the device. The insulator is bored through with tiny drills, and fitted with the germanium crystal and the emitter and collector leads. Rather than rely on the usual cat's whiskers, Helsdon's transistor employs two minuscule, pointed, blade-like electrodes. Fashioned from tiny pieces of phosphor bronze wire, pounded flat, and ground to an arrowhead-like shape, their points rest on the face of the crystal under the mild pressure of their own elasticity. Figure 14.6 summarizes the design.

The Plumber's Point-Contact Transistor

In building my own transistors, I drew upon the lessons of Atkins, Turner and Helsdon. I wanted a layout that was simple to build with contemporary parts, and easily adjustable. At the same time, I wanted the benefits of mechanical rigidity, and protection for the crystal from environmental factors. Finally, I wanted a design that would allow flexibility in the experimental use of different crystals and electrode materials.

I did make several attempts to arrive at a good formula from which another enthusiast could reproduce my work. A few examples can be seen in figure 14.7. The first transistor was made from the parts of an old pressure-relief valve. Later, I combined a handful of common copper and brass plumbing parts to make excellent experimental transistor housings. For obvious reasons, I call the later model, "The Plumber's Point-Contact Transistor." A cutaway view with the major components labeled can be found in figure 14.8. As is so often the case, description of the assembly process is much more complicated than the process itself.

Fabrication began with the transistor's shell. This consists of a 1-inch length of 3/4-inch copper pipe pressed into a 3/4-inch copper

Figure 14.7
To the left and right are examples of the Plumber's Point-Contact Transistor; in the center, an earlier model made from a relief valve

cap. The tube is soldered into place, so that it can't be removed. A second 3/4-inch cap, which will function as a protective cover or lid, can be slid on and will remain in place by friction. See figure 14.9

My transistors, like other experimental models, are based on ceramic 1N34's. If one is dedicated enough to the hunt, samples of this part may still be found in junk electronics from the 1950's or early 60's. As usual, garage sales, flea markets, and thrift stores are good places to start. Ham radio festivals are another. I have purchased dozens of useable diodes through online auctions. (Author's Note: I urge you to exercise good judgement when deciding to scrap old electronics. Always try to take into consideration the possible historic or nostalgic value of vintage gear. There are collectors and afficionados around the world who would be happy to give you a fistful of old diodes and more, if it saves an irreplaceable piece of electronic history from needless destruction. The Internet is an excellent resource with which to research the potential value of so-called "junk.")

As I alluded to earlier, my technique for removal of the germanium amounts to crushing the ceramic housing. This must be done with extreme care, so as not to chip the tiny germanium slab or

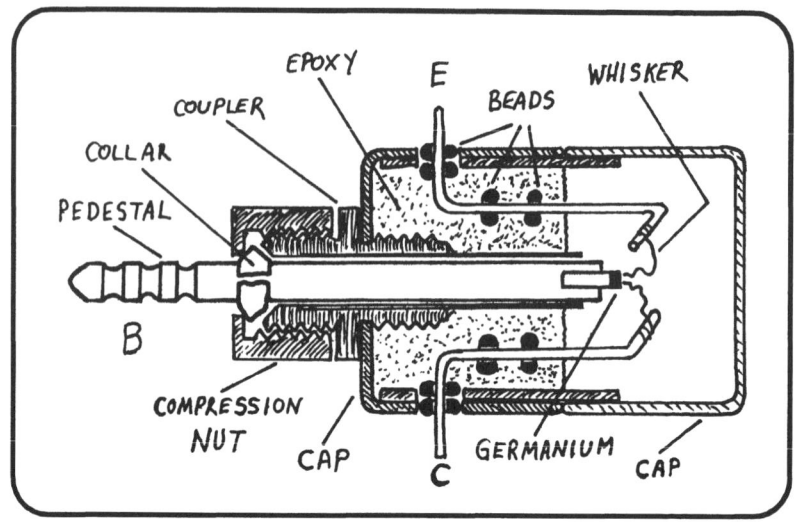

Figure 14.8
A cutaway view of the Plumber's Point-Contact Transistor

die. The die, in most of the cases I've seen, is soldered to a tiny brass pin. Once the diode has been broken open, the pin, with the attached die, can be removed by loosening a tiny set screw and withdrawing it.

Avoid touching the surface of the germanium, and do not allow it to become moist. If the surface is accidentally soiled, one experimenter recommends the use of lacquer thinner as a cleaning agent. If this becomes necessary, wear eye protection, exercise caution, and use it only with adequate ventilation. Helsdon, in contrast, recommends "silicone-impregnated lens tissue" as a cleaning agent.

The pin is mounted in a part I call a pedestal. The pedestal is nothing more than a 2-1/2-inch length of 1/4-inch brass rod. At one end of the rod, I drilled a small hole in the center of face. This hole is sized to accept the brass pin on which the germanium crystal is attached. The exact dimensions of the hole depend on the diode from which your pin has been extracted. Expect to use something on the order of a 1/32-inch bit, to a depth of, say, 1/4 inch. See figure 14.10.

The other end of the pedestal rod requires no treatment at all. However, since it will be later used as a sort of handle by which to adjust the placement of the crystal, I took the liberty to taper it in a pleasing fashion, and to cut shallow grooves or rings to improve the grip of my fingers. I have a small jeweler's lathe that makes operations of this type a snap. Nearly the same effect can be achieved by chucking the pedestal in a hand drill or drill press, spinning it, and dressing the rotating part with a triangular metal file.

The pedestal is held in place by an assembly I call the pedestal-clamp. The pedestal-clamp, seen in figure 14.11, is fashioned from a common brass coupler of the type used to join two lengths of flexible copper tubing. I purchased a 1/4-inch coupler, which comes with two compression nuts and two compression collars.

One end of the coupler has to be extended. This is done with a 1-inch length of 9/32-inch brass tubing. The tubing must be slid into one end of the coupler and soldered into place. Since the inside diameter of the coupler will almost certainly be too small to allow this, you'll have to enlarge the bore with a drill bit. Avoid pushing the drill all the way through the fitting or you'll ruin it for our purposes. You only need to enlarge the bore to a depth of about 3/16 inch.

Ideally, the internal bore of the coupler, including the extension, should be 1/4 inch throughout its length. It's possible that the manufacturer may have restricted the passage inside the coupler somewhere along its length, or a tiny gob of solder may have intruded

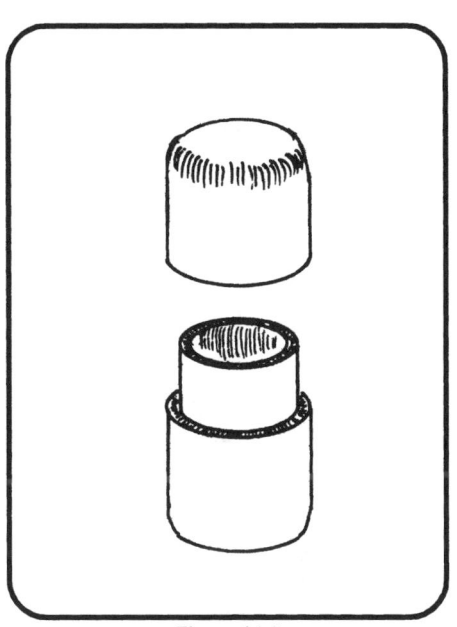

Figure 14.9
Components of the transistor's shell

into the bore when you soldered the brass extension into place. The solution is to run a 1/4-drill bit through the entire bore of the coupler, from one end to the other, to ensure smooth passage.

With the pedestal-clamp complete, it must be installed in the transistor shell. I located the center of the floor of the shell and drilled a 7/16-inch hole all the way through. I inserted the pedestal-clamp assembly through that hole, so that the extension points into the interior of the shell. I soldered the assembly into place from the outside of transistor. See figure 14.12.

Next, I drilled two holes into the side of the transistor to accommodate the emitter and collector electrodes. The holes measured about 5/32 inch in diameter, and were located 180 degrees apart, 1/4 inch up from the floor of the transistor's housing.

The emitter and collector electrodes are "Z"-shaped, and were fashioned from 1/32-inch copper tubing. Tubing of this type is available at hobby stores. Solid copper wire of a similar diameter could probably have been used to equal effect.

If you study figure 14.13, you'll see how the electrodes are fitted into the transistor body. The head of each "Z" is suspended over the extension rising from the pedestal-clamp. The body of the "Z" descends into the transistor housing. The foot of each "Z" projects through its respective hole in the side of the housing and serves as the terminal through which electrical connections will later be made.

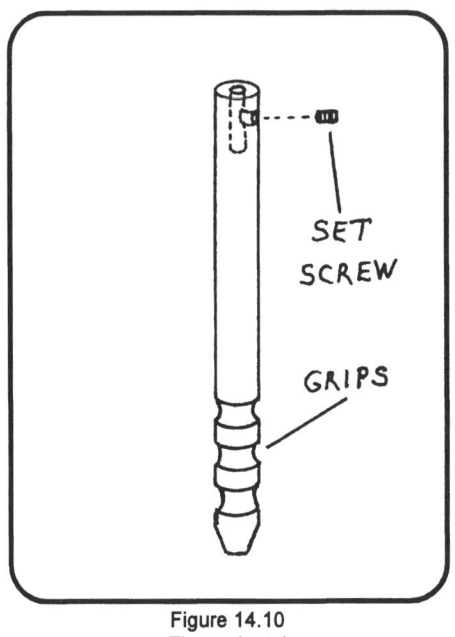

Figure 14.10
The pedestal

The body of the transistor is, of course, conductive. In fact, it is used as an electrical connection to the germanium die itself. It stands to reason, then, that the emitter and collector electrodes cannot be allowed to touch the housing or pedestal-clamp at any point along their length.

To address this concern, I employed glass beads, which were secured from a craft store. Numerous beads were threaded onto the emitter and collector electrodes before they were bent into the "Z"-shape. A tiny drop of hobby cement was used to secure the beads precisely where I wanted them. Figure 14.13 shows how these beads were put to good use. Note that each electrode is centered in the hole where it exits the transistor, and is thereby insulated from it, by a black, glass bead.

The electrodes may be well insulated by the beads, but at this stage, they're not really secured to anything solid. To anchor them and make the transistor, as a whole, rigid and stable, I used a slow-set liquid epoxy. I mixed up the epoxy per the manufacturer's instructions, and poured it into the housing until the level rose to within an 1/8 inch of the open end. I allowed it to cure for a full 24 hours.

The cat's whiskers in The Plumber's Transistor are made from phosphor bronze wire. I have no doubt that such wire is available through mail order, or perhaps on the Web. Luckily, I had a handy source of my own, as I play guitar. My acoustic guitar strings have a steel core, but are made thicker by a wrapping of fine, phosphor bronze wire. Strings grow "stale" and lose their brilliant tone as they're played, so I'm always replacing them. The trimmings from a fresh set of new strings provides enough clean wire for hundreds of transistors. If you're not a player, you might still consider a music store as a source of small quantities of phosphor bronze.

The outside diameter of the guitar string I used was 0.026 inch. The phosphor bronze wrapping, unwound from the steel core, measured 0.007 inch.

When removed from the steel core, the bronze wire has a pronounced ripple to it, which can be relieved by placing it under tension and raking it across the edge of a screwdriver. I placed a sample of the straightened wire on a piece of smooth steel and used a small hammer to flatten it. The resulting ribbon was probably half the original wire's thickness.

Two pieces, cut from the bronze ribbon, must each be sharpened to as fine a point as possible. A strip of emery paper glued to a popsicle stick makes an effective file. Once the whiskers were sharpened, I used tweezers and surgical forceps to bend them into their required "L"-shape. In this case, each leg of the "L" measured 1/4 inch, or just under. Eye loupes are inexpensive, but indispensable for inspecting your work and gauging your progress. A loupe with 4X magnification is perfect.

Next, the whiskers had to be attached to the emitter and collector electrodes. Figure 14.14 shows the basic idea. Using a hot iron and a dot of solder, each whisker was tacked to its respective electrode wire. The whiskers were oriented so that their sharpened points were centered over the open mouth of the pedestal-clamp's extension tube.

Using tweezers, I picked up the brass pin and germanium crystal, previously removed from the 1N34 diode, and inserted it in the hole at the end of the pedestal rod. If you're lucky, friction will hold the pin in place. Solder may be an option, though in my transistors, I took the more sophisticated route of installing a tiny set screw. I inserted the crystal pin, tightened the screw, and locked it into place.

The pedestal, with the attached crystal, was inserted into the pedestal-clamp, and pushed halfway into the body of the transistor. I threaded the compression collar onto the rod, followed by the compression nut, and tightened it lightly by hand.

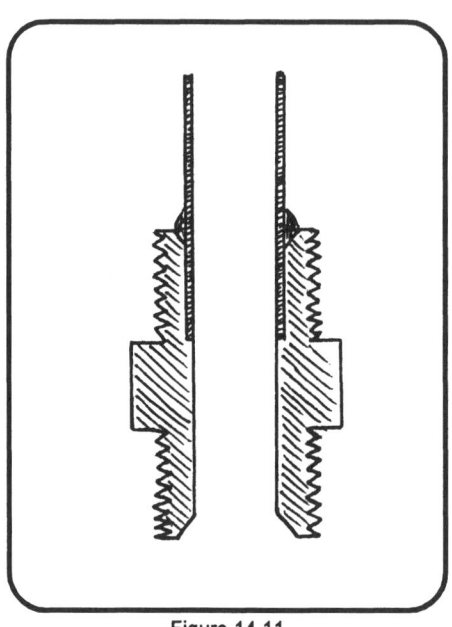

I should mention that plain compression collars are too rigid for this application. It takes far too much torque on the compression nut to cause the fitting to grip the pedestal rod. Often, when it finally does, the act of tightening the nut causes the pedestal itself to move. This is not good.

A simple modification takes care of the problem. Using a hobby grinder fitted with a small cutting wheel, I cut through the collar at one spot, changing its closed "O"-shape into a nearly closed "C". The modified collar grabs the pedestal rod securely, requiring little torque on the compression nut to do it. It also cured the problem of "creep" in the rod as the nut was tightened.

Figure 14.11
The pedestal clamp

The final assembly step involved the adjustment of both the pedestal rod and the cat's whiskers. The whiskers must be tweaked so that their points rest firmly on the face of the germanium crystal, roughly normal to its surface. The whiskers must lie about 0.002 inch apart, and cannot be allowed to touch each other. By loosening the compression nut on the pedestal-clamp and sliding the pedestal rod inward, I was able to bring the crystal into contact with the whiskers, and adjust the relative pressure under which the germanium is probed.

When assembly was complete, and all adjustments had been made to my satisfaction, I popped the cap on the transistor housing and closed it up.

While this completes the mechanical assembly, the transistor must be electrically tested, characterized, and if necessary, adjusted. This is most easily accomplished with Turner's test setup, reproduced in figure 14.15. Basically, the circuit requires two ammeters, two switches, and two batteries.

One battery is 1.5 volts. A single common flashlight cell will suffice. The second battery is a 45-volt supply. Five 9-volt batteries, wired in series, will produce the necessary voltage.

To begin the characterization, we must first identify the transistor's terminals. The copper housing, which provides a conductive path to the germanium die, becomes the base. The emitter and collector electrodes, however, are not yet identified in any conclusive way. What I mean to say is that the emitter electrode may, in truth, be the collector, and vice versa. Therefore, for the first stab at characterization, the choice of which is which is purely arbitrary.

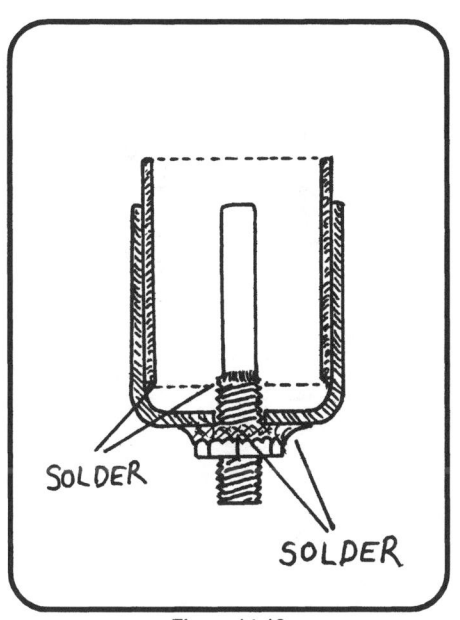

Figure 14.12
Transistor shell and pedestal clamp

Before engaging in Turner's process, I like to use an ohmmeter to measure the resistance between the emitter and collector electrodes. If this value is low, it probably means that your cat's whiskers are touching each other. Readjust and retest the whiskers before you proceed.

Assuming that the emitter and collector whiskers are properly spaced, the first step in the testing process is to close S_1, leaving S_2

open. Refer to figure 14.15. Turner reports that the emitter current, as indicated by M_1, should be on the order of 0.02 amp or less.

Next, S_1 is opened. S_2 is closed, and the collector current is measured by M_2. Turner states that it should not exceed 0.0005 amp. This value can be used later to help quantify the transistor's ability to amplify, so it should be recorded on a sheet of paper with the designation "I_1".

In the real world, one of the above measurements are often too high. If either the emitter or collector current is in excess of the suggested values, the first trick is to swap the emitter and collector terminals and repeat the two tests. If the measured values suddenly fall into the appropriate ranges, this means that your initial guess at which electrode was the emitter and which the collector was in error. I used a dot of ink to permanently mark and identify the correct emitter and collector terminals.

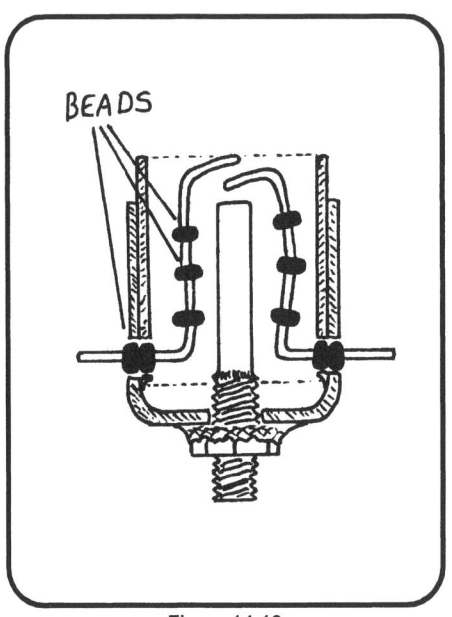

In the event an emitter/collector swap fails to address an over-current condition, the problem may lie with the regions of the die on which the cat's whiskers rest. The solution is to loosen the pedestal-clamp, draw the crystal away from the whiskers, rotate it a few degrees, and then bring the crystal back into contact with the whiskers. This will present a fresh surface to the electrodes. Of course, it becomes necessary to repeat all of the prior electrical tests.

Figure 14.13
Using glass beads as insulators to prevent electrode wires from contacting the pedestal clamp and wall of the transistor's shell

Assuming all has gone well, the final test in Turner's process is to close S_1, leaving S_2 closed from the previous step. The reading on M_2 will probably rise to a new value, which should also be recorded, and designated "I_2".

End Results and Going Further

In judging the merit of each of the amplifiers we've explored thus far, our practice has been to compare the amplifier's output to its input. By dividing the latter into the former, we arrived at a ratio which describes the amplifier's ability to regenerate and magnify input signals. As you recall, we referred to that as gain.

A simple gain calculation presupposes that the type of measurements made to the output side are the same as the those made to the input, in order that the comparison be meaningful, i.e., volts to volts, amps to amps, or whatever.

When we considered the Votive Triode, we noted that the output was expressed in terms of current, while the input was measured in volts. We can't use these numbers to compute gain in the usual sense.

The ratio of output current to input voltage was called *mutual conductance*. It wasn't the same thing as gain, though we demonstrated that gain could be computed on the basis of it.

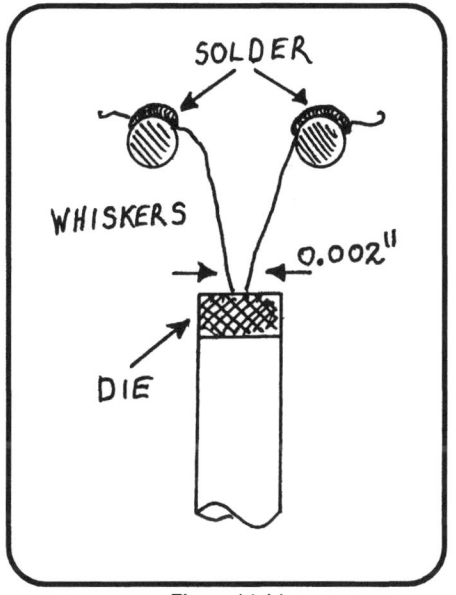

Figure 14.14
Installation and placement of whiskers

To an extent, Turner's hookup mimics the conditions we saw with the thermionic triode. The output of the transistor, like the triode, is clearly expressed as a change in current. The input to Turner's setup is a 1.5-volt battery, which is either on or off. The input for Turner's configuration, like the input for the triode, is a voltage. So, like the triode, we can use current and voltage values to compute mutual conductance. From mutual conductance, we may later compute gain.

To compute mutual conductance, Turner used the equation in figure 14.16. The phrase "(I_2-I_1)" represents the change in collector current as the emitter battery is switched on and off. The "1.5" value reflects the change in emitter voltage as S1 is opened and closed. Obviously, the emitter voltage is 1.5 volts when the switch is on, and 0.0 volts when it's off. The change in voltage is thus (1.5 - 0.0), or 1.5 volts in the denominator.

You may have noted the term *transconductance*, a phrase Turner uses in lieu of *mutual conductance*. In reality, the two mean precisely the same thing, it's just that the latter is now more fashionable.

Transconductance actually has a unit associated with it, the inverse ohm, or *mho*. The higher the transconductance value, the greater the transistor's potential as an instrument of amplification. Turner claimed to have achieved transconductance values as high as 5,000 micromhos (millionths of a mho), with 1,000 to 3,000 being typical. Data from a construction article by C. E. Atkins implies his best came in at about 3,000. My personal experience in this regard has been mixed. I've built a fair number of nonfunctional transistors, and a few that reached 2,800 micromhos.

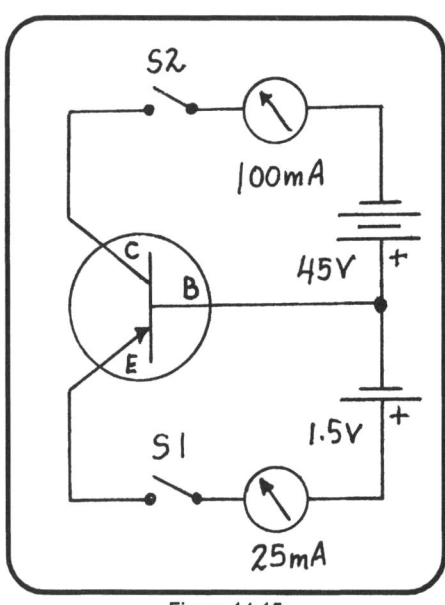

Figure 14.15
Turner's test circuit

Like any other transistors, the point-contact devices made by this and similar methods have operating parameters that must be considered. If voltages or currents are applied to it in excess of those limits, the transistor is likely to be damaged. The good news is that, since it's a homemade device, it can be dismantled, readjusted, and placed back into service. A given piece of germanium will suffer a fair degree of abuse before it's so degraded that no useable surfaces are left for your use.

Nonetheless, Helsdon quotes some values that are worth keeping in mind as you tinker. DC voltages, applied to the collector, should not be greater than -30 volts with respect to the base. AC peak voltage should not exceed -80 volts. The collector current should be less than -10 milliamps, and the maximum power dissipated by the collector should be limited to something under 50 milliwatts.

Noteworthy results can be obtained following the guidelines above, though it's possible to enhance the amplifying qualities of the transistor even further. This requires special treatment of the collector-base junction with a technique known as *forming*.

Forming is largely a thermal process. In a nutshell, a brief high-current pulse is delivered to the junction, which causes localized heating. The heating contributes to the development of the P-type region under the collector whisker. Once again, the references I have differ in their explanation of what, precisely, occurs. Some imply that the heat causes changes in the N-type crystal. Others suggest that, in the case of phosphor bronze whiskers, phosphorus atoms actually migrate into the germanium, thereby doping it. In any event, there's no disagreement that forming improves the point-contact transistor's ability to amplify.

Forming is a tricky process, where carelessness can lead to complete destruction. Figure 14.17 shows the

$$\mu\mho = \frac{1000 \times (I_2 - I_1)}{1.5}$$

Figure 14.16
Turner computes transconductance in micromhos

schematic for a forming rig composed of a switch, a capacitor, a resistor, and a power supply consisting of a stack of 9-volt batteries. When the switch is in the "Charge" position, current is applied to the capacitor through the current-limiting resistor, "R." The capacitor will accumulate a charge until its voltage matches the voltage of the power supply. When the switch is thrown into the "Form" position, the charge stored in the capacitor is rapidly dumped into the collector-base junction. The forming process halts when the capacitor is completely discharged.

Helsdon's regimen for forming junctions is attributed to an *R.C.A. Review* article by B. N. Slade entitled "Factors in the Design of Point-Contact Transistors" (1953). I haven't seen the article myself, but Helsdon implies a circuit similar to figure 14.17, in which the chosen capacitor is 0.001 microfarads. The power supply is set to 80 volts or so. The capacitor is repeatedly charged and then dumped into the collector-base junction. Each time capacitor is recharged, the supply voltage is raised another 20 volts. The total supply voltage is eventually raised to 300 volts.

Once this sequence of voltages is completed, the capacitor is replaced with a larger value, say 0.005 microfarads. The forming voltage is returned to 80 volts, and the entire sequence is repeated. As each subsequent cycle is completed, the capacitor value is increased, until it reaches a value of 0.1 microfarads.

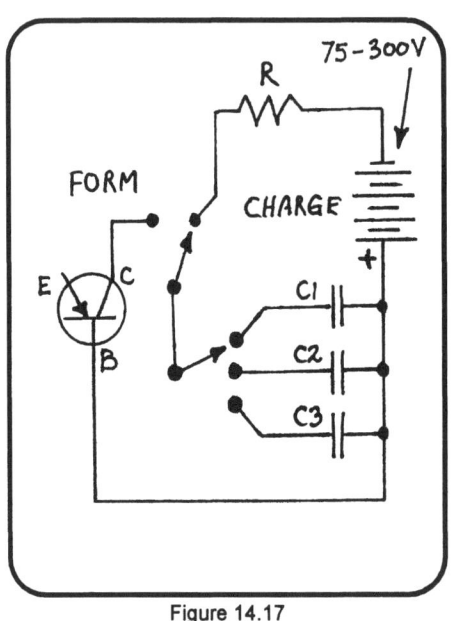

Figure 14.17
A junction forming rig

As I've already said, it's quite easy to destroy the junction in an attempt to improve it, so it pays to monitor your progress as the forming process continues. After every application of current to the junction, Helsdon suggests testing it with an ohmmeter. The forming process should be halted when the junction resistance is less than 30,000 ohms with no voltage applied to the emitter.

So what can you do with the Plumber's Point-Contact Transistor? Apparently, quite a bit. Helsdon's article discusses the application of homemade point-contact transistors, similar to mine, to sawtooth generators, audio oscillators, receiver circuits, and audio amplifiers. He cites a single-stage amp, following a simple crystal receiver, delivering a power gain of 20 to 1.

Atkin's article actually included two schematics, one for an experimental amplifier circuit, and one for an oscillator. Both were based on his homemade transistors. Reportedly, the amplifier produced a power gain of 10 to 1. The oscillator generated alternating currents with an amplitude of 15 volts, ranging in frequency from 20 to

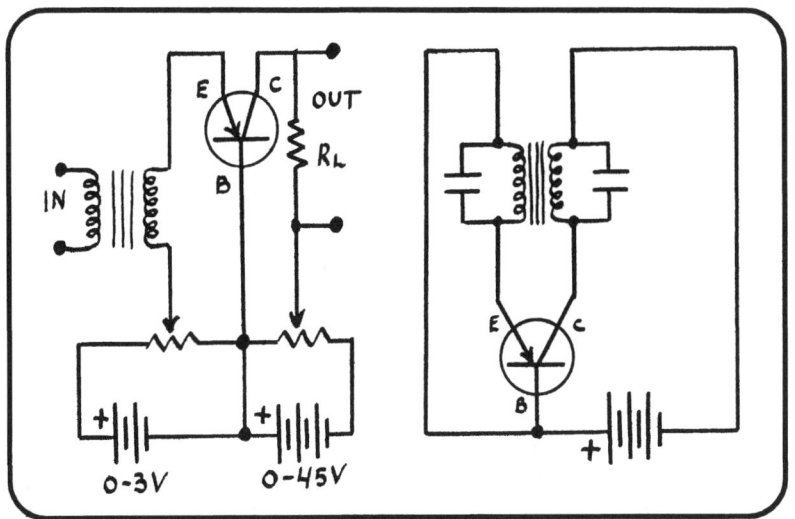

Figure 14.18
Point-contact transistor circuits—an amplifier and an oscillator

100 kHz. These circuits are represented in figure 14.18.

Still More Ideas

Phosphor bronze seems to be the preferred material for cat's whiskers, being mentioned in numerous books and other references. On the other hand, both Turner and Atkins demonstrated that tungsten performs equally well. It was interesting to find at least one reference to beryllium copper wire, used specifically for the emitter whisker. It's unclear to me what advantage this material is supposed to have offered, but if you're interested, it's easy enough to obtain and experiment with.

The quantity of copper beryllium wire you need to build a transistor is so small that it probably doesn't warrant purchasing a spool of it, presuming you can find somebody who sells it. Salvage may be a better option, if you know where to look. The cabinet panels on old computer or communications equipment sometimes feature springy metal "fingers" or tines, which are intended to provide electrical continuity between surfaces when the cabinet is assembled. If they're thin, copper colored, but highly elastic, the tines might be fashioned from beryllium copper. Another place to look is inside old electronic

Page 249

connectors, sockets, or switches. Again, look for small metallic parts that appear to be copper, but are very flexible. It has occurred to me that old, broken panel meters may be an excellent source of whisker material. Most meters I've seen contain a spiral-wound hairspring, which serves the dual purpose of passing current to the coil that moves the needle, and returning the needle to the zero position when the current is removed. While I have not confirmed this, I strongly suspect that the springs in at least some of the meters I've disassembled are made of a beryllium copper alloy.

I've already discussed the process of cutting, shaping, and sharpening the cat's whiskers. The same will have to be done to any sample of beryllium copper you happen find, which means you'll generate shavings, filings, and dust. You should be aware that beryllium, in its pure form, is horribly destructive to lung tissue. Knowing this, I searched the Web looking for an MSDS sheet on copper alloys containing beryllium. My conclusion is that there's little risk using it for the purposes described in this book, though I'd probably wear a dust mask, and perform any shaping, sanding, or soldering operations outdoors where there's plenty of fresh air. (Undoubtedly, I'm being overly cautious, but I see no harm in that. In general, I advise all of my readers to inspect the MSDS sheets on any material they might use. Please render your own decisions.)

The ceramic package 1N34, on which the Plumber's Point-Contact Transistor is based, has long since become obsolete. One would think them difficult to find, though I've had little difficulty locating pieces here and there. In fact, even "burned out" samples of the vintage 1N34's, useless to anyone else, are just fine for transistor construction purposes.

Of course, the supply of available vintage 1N34's will eventually dry up. While you're looking for them, keep an eye out for 1N21's, 1N60's, CG1's, CG4's, Transitrons, or other diodes from that period. I've even played a bit with modern germanium diodes.

Sadly, modern diodes are more difficult to work with, because they're sealed in clear glass beads, which must be crushed to extract the germanium. Needless to say, the crystal is at risk throughout the procedure. As well, diode manufacturing has become much more efficient, so the amount of germanium used per diode has shrunk to the smallest possible quantity. This means that crystals removed from modern diodes make exceedingly small targets on which to align two cat's whiskers. I actually purchased a stereo microscope to aid me in manipulating tiny parts, and have found it indispensable. In addition, Helsdon suggests that more modern diodes may have been "treated" to processes that enhance their performance as diodes, but make them unsuitable for use in experimental transistors. If your transistor fails to work properly, you may need a different sample of germanium.

References

Coblenz, Abraham, and Owens, Harry L., *Transistors: Theory and Applications*, copyright 1955, McGraw-Hill Book Company, Inc., New York, pp. 58-72, (point-contact transistors)

Conti, Theodore, *Metallic Rectifiers and Crystal Diodes*, copyright 1958, John F. Rider Publisher, Inc., New York, pp. 4.8-4.9, (rectification in the crystal diode)

Dewitt, David and Rossoff, Arthur L., *Transistor Electronics*, copyright 1957, McGraw-Hill Book Company, New York, pp. 337-338, (point-contact transistor details)

Hunter, Lloyd P., Editor, *Handbook of Semiconductor Electronics*, copyright 1956, McGraw Hill Book Company, Inc., New York, pp. 1.1-1.3, (point-contact transistor characteristics), pp. 4.17-4.20, 4.27, (details of point-contact transistor construction and operation), pp. 5.3-5.8 (photoconductivity in point-contact devices), pp. 8.7-8.8, (metal-semiconductor contacts, and forming details), p. 9.7, (encapsulation and effects of moisture on junctions)

Kiver, Milton S., *Transistors*, copyright 1962, McGraw-Hill Book Company, Inc., New York, pp. 4-30, (semiconductor physics), pp. 30-46, (junction behavior), pp. 46-50, (point-contact transistor detail)

Krugman, Leonard, *Fundamentals of Transistors*, copyright 1954, John F. Rider Publisher, Inc., New York, pp. 9-11, (operation of point-contact transistors)

Markus, Abraham, *Radio Servicing Theory and Practice*, copyright 1960, Prentice-Hall, Inc., Englewood Cliffs, New Jersey, pp. 150-160, (crystal diodes and triodes, point-contract transistors)

Pollack, Harvey, *Transistor Theory & Circuits Made Simple*, copyright 1958, American Electronics Co., Mineola, L.I., New York, pp. 23-28 (operation of point-contact transistors)

"A Crystal That Amplifies - Details of Experiments With a Double-Contact Germanium-Type Crystal Having Transconductance," C. E. Atkins, *Radio & Television*, October, 1948, Ziff-Davis Publishing Company, pp. 39, 181-184

"Home-Made Transistors - Inexpensive Conversion of Selected Germanium Diodes," P.B. Helsdon, *Wireless World*, January, 1954, pp. 20-23, (details on constructing a homebrew point-contact transistor)

"Build a Transistor," Rufus P. Turner, Radio Electronics, May, 1949, pp. 38-39, (details on constructing a homebrew point-contact transistor)

Chapter 15
The Cuprous Oxide Transistor

Without doubt, the invention of the point-contact transistor marked the beginning of a new age in electronics. Yet, despite the real advantages that it offered over comparable thermionic amplifying devices, in the end, it represented a fairly short-lived technology. Point-contact transistors, like vacuum tubes, have shortcomings of their very own.

If you've endeavored to recreate the experiments in the last chapter, you're already well aware that aligning cat's whiskers onto the face of a tiny crystal is a delicate business. It's time and labor intensive, and not something that lends itself well to mass production. Affordable technologies demand automation and require the economy associated with volume production.

Once proper assembly and alignment has been achieved, it's quite possible to disrupt the positioning of the whiskers, and thereby ruin the transistor, by subjecting it to shock or by applying heat or torsional forces to its terminals.

Point-contact transistors have significant electrical limitations, too. As amplifiers, they're noisy devices, so their usefulness in real high-gain applications is limited. In addition, point-contact devices are not capable of dissipating large amounts of power. Consequently, they're suitable for use only in low-power applications.

Junction Transistor Basics

Point-contact devices may have ushered in a new age, but the real revolution in solid-state electronics came with the introduction of junction transistors. Junction transistors are much more quiet in operation, they can be engineered to handle relatively large amounts of power, and ultimately, they can be mass produced for throwaway prices.

While junction transistors have appeared in countless forms, the product of an endless variety of manufacturing processes, they all share a basic internal structure. Junction transistors don't rely on cat's whiskers, but are instead built of alternating layers of semiconducting materials like the junction diode described a few chapters back.

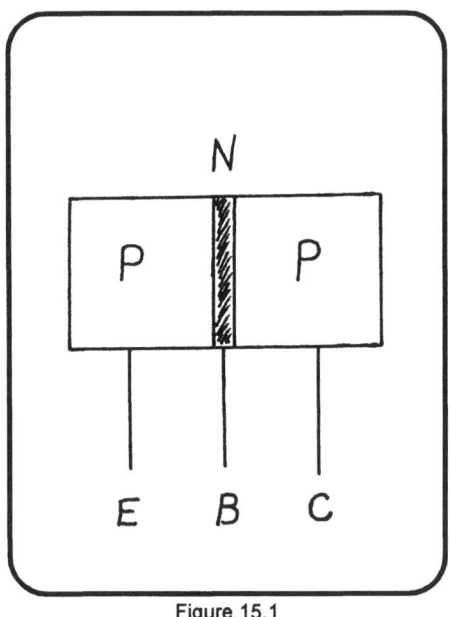

Figure 15.1
A PNP junction transistor

Whereas the diode consisted of two layers of semiconductor material, an N-type and a P-type, the junction transistor is a triode. It's built from *three* layers of semiconductor material. Figure 15.1 illustrates a typical junction transistor composed of some N-type material bound on either side by layers of P-type material. The leftmost layer, made of P-type material, is the emitter for this particular transistor. The middle section, composed of N-type material, serves as the transistor's base. The P-type section at the far right of the sandwich is the transistor's collector. A transistor of this type is referred to as a PNP transistor. The "PNP" nomenclature gives us an indication of the type and arrangement of semiconducting materials in the transistor. (For the record, it's possible to shuffle things around to produce an NPN transistor. In fact, most modern junction transistors are NPN devices. For the sake of uniformity, we'll stick with the PNP devices for the moment.)

The specific details behind the operation of the PNP junction transistors differ somewhat from those of the point-contact transistor, but the general idea is the same. As we'd expect, in normal use the collector-base junction is reverse-biased. This means that by default, no charge carriers are available to cross the collector-base junction, so no collector current can flow. The transistor is considered to be switched "off."

When a positive voltage is applied to the emitter, the emitter-base junction is forward-biased. As a result, holes are injected into the emitter's P-type material, where they rapidly migrate toward the P-N junction formed by the emitter and base.

The holes cross the junction into the base's N-material. The base is purposefully engineered to be physically very narrow—so narrow, in fact, that most of the holes have no problem wandering through the base right into the collector. (The collector, of course, is negatively charged, so it happily attracts and welcomes those holes.)

The appearance of charge carriers in the collector (in the guise of injected holes) allows a current to flow through the collector-base junction. The transistor is now "on."

At the risk of being repetitive, we can say that the device amplifies because the magnitude of the current in the collector circuit can be controlled by adjusting the quantity of holes injected into the emitter.

While we're on the subject, note that the dominant or *majority* charge carrier in the PNP devices is the hole. In the NPN transistor, where the semiconductor types have been reversed, electrons are the majority carrier.

The Cuprous Oxide Transistor

Homebrew construction of junction transistors does not appear to be feasible, at least not the fabrication of germanium devices. Unlike the case with point-contact transistors, I was unable to find any references to the homebrew fabrication of junction transistors. I fear that the nature of the tools, materials, and techniques probably lies well outside of the scope of all but the most fanatic of amateur scientists.

Just the same, in researching the subject, I began to daydream about alternative materials that might be used for transistor construction—materials more commonly available, or more easily created.

In the 1920's, L.O. Grondahl and P.H. Geiger introduced a diode composed of a copper plate coated with a thin layer of cuprous oxide. The plate formed the negative terminal; the coating, a positive terminal. These junctions could be manufactured in large sizes, which translated into the useful ability to handle large currents.

It occurred to me that it might be possible to arrange copper and oxide layers into the now-familiar PNP sandwich, thereby producing (I hoped) an instrument of amplification.

I read, pondered, and tinkered. In the end, my preliminary experiments didn't result in what I'd consider a practical amplifier. In fact, calling it a "transistor" may represent some artistic license on my part. On the other hand, the results were encouraging enough that they probably warrant further experimentation. Let me show you what I did.

Producing a Layer of Oxide

The first step toward testing my idea was to learn how to create the necessary oxide layer on a copper surface. The process is not overly difficult, but it was not as easy as I had first imagined.

To begin with, copper can combine with oxygen to produce two different forms of oxide. The first is *cupric oxide*, formula CuO. As you can see, each molecule is composed of a single copper atom combined with a single atom of oxygen. This substance is black in appearance. It's unuseable for our transistor application.

Another form of copper oxide is called *cuprous oxide*, formula Cu_2O. Note that the cuprous oxide molecule contains two atoms of copper for every atom of oxygen. It has a deep salmon color. This is the material from which copper/copper oxide PN junctions can be created.

Preliminary research on the subject turned up a booklet written by Walt Noon, entitled *How to Build a Solar Cell That Really Works*. In it, Noon describes a process for building a cuprous oxide–based solar cell. As you may or may not know, solar cells are simply PN junctions that can produce a voltage on exposure to light.

Mr. Noon's process for producing a cuprous oxide layer is simple. He starts with a copper sheet, which is vigorously scoured to remove any visible contaminants. The copper is then etched in a nitric acid bath composed of 20 parts nitric acid to 80 parts distilled water. (Warning: Always add acid to water, not the other way around!)

The cleaned sheet is then heated over the flame of a Bunsen burner or propane torch. The copper is heated on one side only, until it's red hot. It's kept at the red-hot temperature for just under three minutes.

The plate is then removed and allowed to cool slowly. By the end of the process, the unheated side of the plate has become blackened. This black pigment is cupric oxide. Just beneath it, however, is the desired layer of cuprous oxide. To remove the cupric oxide, while leaving the cuprous layer intact, Noon immerses the plate in nitric acid once more. This eats away the black oxide. The plate is removed, rinsed, and buffed lightly.

Obviously, completing Noon's solar cell requires additional steps, but those weren't relevant to my objective. I thought I understood the oxide process well enough to attempt to build my first transistor.

A First Attempt to Create a PNP Structure

My idea for a homebrew junction transistor was a PNP device comprised of a metal copper base, coated on either side with cuprous oxide. One coating would become the emitter; the other side, the collector.

The base of my transistor was fashioned from a piece of clean, soft copper sheet with a nominal thickness of 0.025 inch. Using tin snips, I cut out a piece about an inch long and a half-inch wide.

While 0.025-inch copper makes for a nice, rigid substrate on which to build a transistor, the transistor's operation depends upon the migration of charge carriers through the base. Given what I'd learned about electrode spacing in the point-contact transistor, I suspected that my base material might be far too thick.

To remedy this, I hit upon the idea of dimpling the copper strip from either

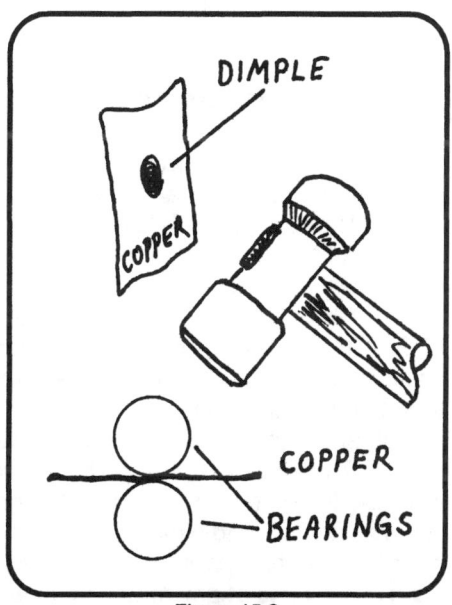

Figure 15.2
Dimpling the copper sheet to create a thin base region

side to create a local region in the copper where it would be exceedingly thin.

Dimpling can be done with a hammer and a pair of chrome ball bearings. I placed the first bearing upon the surface of a piece of scrap wood. I set the copper strip on top of it, then the second bearing above that. I struck the entire stack with the swift blow of a hammer. The result is depicted in figure 15.2. The thinner the layer separating the two dimples, the better off you are. Make sure you wear eye protection in case one of those bearings zips out sideways and launches for the moon.

Next, I tried to coat both sides of the dimpled strip with cuprous oxide. I followed Noon's process as well as I could.

For testing purposes, each junction in the transistor, the emitter-to-base, and collector-to-base, can be thought of as a separate diode. With a digital multimeter, I tried to assess the condition of the coatings, measuring conductivity under forward-biased and reverse-biased conditions. To my dismay, all readings indicated a direct short circuit. The process had failed!

I repeated the procedure again and again, each time testing the end results. Only one in thirty-two attempts yielded a single PN junction on even one side of the base. I was never able to create a useable oxide later on *both* sides.

A Second Attempt to Create a PNP Structure

Back at the drawing board, I came upon a reference to cuprous oxide in Martin and Hill's *Manual of Vacuum Practice*. Without going into excruciating detail, suffice it to say that it and a great deal of subsequent trial and error resulted in a modified coating process that now works very well.

My process, like Mr. Noon's, begins with mechanical cleaning of the dimpled copper substrate. Scouring is okay, but avoid leaving significant scratches on the surface of the metal. My copper strips were then bathed in a strong solution of lye (drain cleaner) to remove oils and other organic residue. The strip was well rinsed in distilled water.

The strip was then etched in acid, followed by another rinse with distilled water. In my case, I used *hydrochloric acid* instead of Noon's nitric. I found nitric acid somewhat difficult to come by. I know of no local sources, while hydrochloric acid (commonly called muriatic acid) is available at any hardware store or seller of pool maintenance supplies.

(Consider this too: A good reason to use locally available chemicals in lieu of more exotic substances is rooted in our nation's "war on drugs" and now the recent terrorist attacks suffered in New York City and elsewhere. You can order nitric acid and other chemicals through the mail, but for better or worse, sales to individuals have come under a great deal of scrutiny. Some supply houses won't deal with individuals at all. The danger, of course, is that the innocent purchase of chemicals or equipment by the science enthusiast may be interpreted as the purchase of raw materials for a bomb or bathtub pharmaceutical. Big Brother *is* watching.)

The next step in my process is to heat the copper. My technique relies on an electric hotplate or stove burner whose coils are set to a dull red glow. I stopped using propane when I realized that the products of combustion in the flame can contaminate and ruin my oxide coatings.

Using clean tweezers, the copper strips are placed on the hotplate, but only for a brief moment. The copper is quickly removed the instant that its shiny surface begins to take on a faint blueish tint or hue.

The copper is then plunged into a saturated solution of *sodium tetraborate decahydrate*, commonly referred to as *borax*. Look for it in your local grocery store among the laundry detergents. To make the necessary solution, take a clean beaker or mason jar, fill it with distilled water, and then begin adding borax. Stir the mixture to help the powder dissolve. Continue to add more borax until the solution will hold no more. At that point, crystals will remain on the floor of the jar no matter how much the mixture is stirred.

Following the borax dip, which need only last an instant, the copper strip is returned to the hot plate, and this time heated to a dull red glow. Leave it there a minute or two, then remove it and immerse it in distilled water.

At this stage, the copper should be well coated with black cupric oxide. Using tweezers, dip the copper into a beaker of hydrochloric acid, immediately followed by immersion in distilled water. Repeat the acid dip, then the water. After two or three iterations, the black oxide will melt away, leaving a beautiful, dark, thick layer of cuprous oxide. Don't be too aggressive with the acid dip, as it's quite possible to etch away and damage the cuprous oxide as well.

Using my revised technique, and testing the PN junctions with my meter, I found that one in two of the processed copper strips had at least one good PN junction on it. About one in ten had useable junctions on *both* sides. This, of course, was the objective.

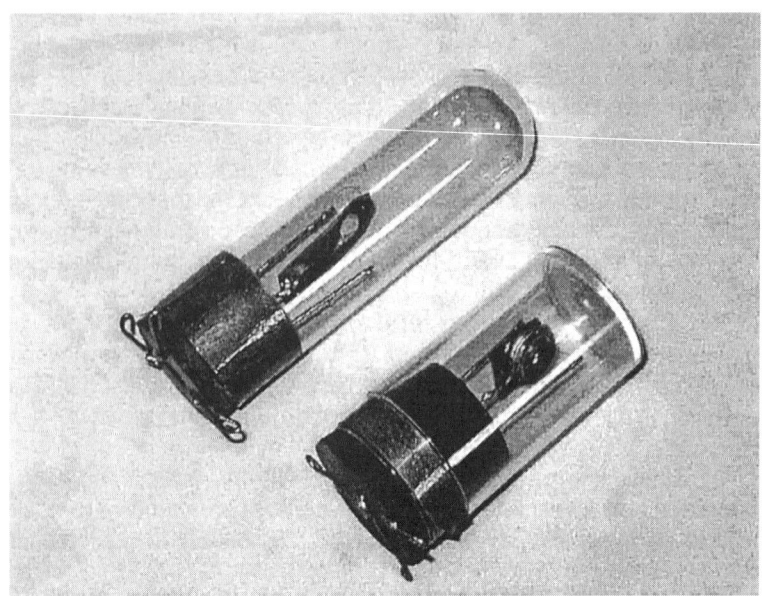

Figure 15.3
Experimental cuprous oxide transistors,
Model I on the left, and Model II on the right

I have another trick worth mentioning. If the copper base strip is made just a little longer than necessary, the added length can be bent over to form an "L" or open "V"-shape. This offers two advantages. The foot of the "L" can be used as a handle by which to manipulate the copper during processing. This minimizes the potential for contaminating the surfaces you intend to oxidize.

Second, the foot will allow you to set the copper strip on edge when placed on the hot plate. Heating the copper on edge assures that neither face will be contaminated by direct contact with impurities on the surface of the hot plate. Later, when all of your processing is complete, the excess length of copper, the foot if you will, can be snipped off.

Building Model I

Once I had a good PNP structure fabricated, the remaining work in constructing my transistor was mostly mechanical in nature. The end result can be seen in figure 15.3.

The foundation for my transistor was a black, rubber stopper, about 7/8 inch in diameter. The stopper fit into a short glass test-tube–like vial, which had previously served as a water bottle for a hamster. The glass tube makes a suitable enclosure to protect the device from dust, moisture, and mechanical damage.

Three pieces of 1/32-inch brass rod were forced through the stopper. One passed through the approximate center of the stopper, and the other two were placed equally to either side. These rods form the terminals for the finished transistor.

To keep the brass rods locked in the rubber and free from unintended movement (particularly rotation), I fabricated some small hook-like pieces which were inserted in the rubber near each electrode, and then soldered to them. Figure 15.4 clarifies this detail.

The center electrode not only serves as the transistor's base connection, but also the mechanical support for the PNP structure I had created earlier. To prepare the copper strip for mounting, I used a pick to

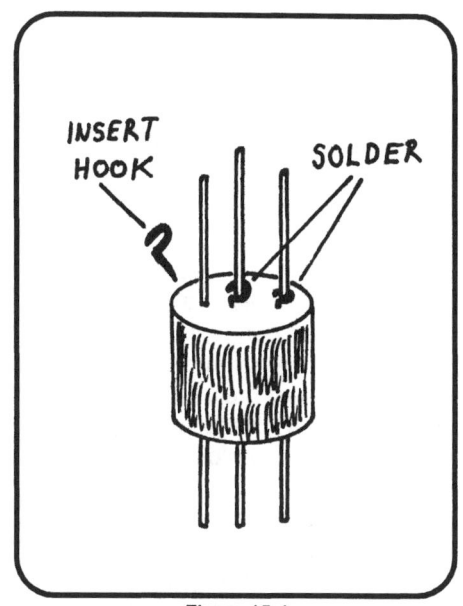

Figure 15.4
The transistor's foundation and electrodes

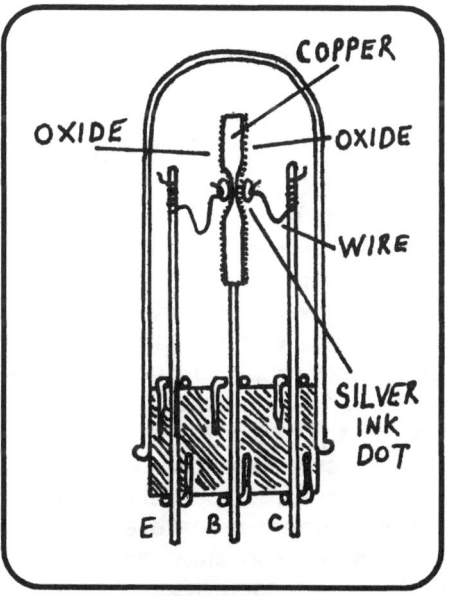

Figure 15.5
A cutaway view of the completed device

Page 263

scrape away some oxide near one end. I tinned that area with a shiny dot of solder, and then soldered the strip to the center electrode.

The last step toward completing the transistor is to connect the emitter and collector electrodes to their respective P-type layers. While the method for making the connection is simple, the effort I expended to arrive at the solution was not. Before you continue reading, I challenge you to consider, for just a moment, how you might go about it.

In the end, I made the connections using 0.003-inch copper wire. I harvested these hairlike conductors by picking individual strands from a thicker piece of flexible, stranded wire.

On the emitter side, I wrapped a few turns of the fine wire around the tip of the brass emitter electrode and then soldered the connection. I terminated the free end of the strand with a tiny loop, and then, using a toothpick, I guided that loop to a dimple in the PNP structure.

The copper strand was actually bonded to the cuprous oxide layer with a tiny drop of silver-bearing ink, applied with another toothpick. Silver-bearing ink is a conductive fluid used for repairing circuit boards. The ink is sold in a special applicator pen. My ink was manufactured by Chemtronics.

When the ink dot was dry, I repeated the process on the collector side. Finally, I inserted the transistor into its protective glass tube. Figure 15.5 is an exaggerated cutaway view of the complete transistor assembly that should answer any lingering questions.

Building Model II

The mechanical construction of the Model II cuprous oxide transistor is similar to that of the Model I, including the rubber stopper, electrode wires, and vial. The Model II differs from the Model I in the form of copper used to fabricate the PNP sandwich.

In the Model I version of my cuprous oxide transistor, I dimpled a 0.025-inch copper strip as a means by which to reduce the thickness of the transistor's base region. The idea is a good one, though I discovered some problems with it.

I suspected that the results of dimpling were probably not uniform from one copper strip to the next. Put another way, it's next to impossible to ensure that every swing of the hammer is identical. Since I didn't have a convenient means by which to measure the end result, the process became inherently uncontrollable.

In addition, like many soft metals, copper hardens and becomes more brittle as it is worked. You may have observed, for example, that a soft, bendable length of paperclip wire will eventually stiffen and fracture if repeatedly bent and straightened.

Under the microscope, there's evidence of tiny cracks in the floor of the dimples, which I attribute to localized hardening as the metal is deformed under the pressure of the ball bearings. The effect of these cracks was to cause a short circuit between the emitter and collector of the transistor, thus ruining the structure.

I contemplated how I might eliminate the dimpling process. The only way, I concluded, was to find copper sheet that was already the desired thickness. This, I found in the form of copper foil, measuring 0.002 inch. Copper foil is frequently used as a shielding material to protect delicate electronics from the influence of external noise sources. It is readily available from a variety of sources.

I cut strips from the

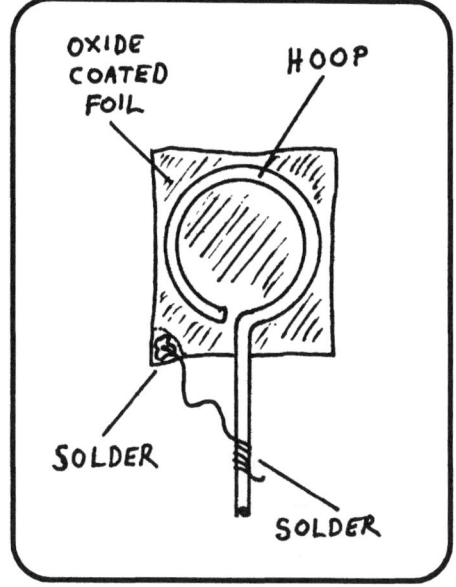

Figure 15.6
Details of the copper foil base and its support hoop

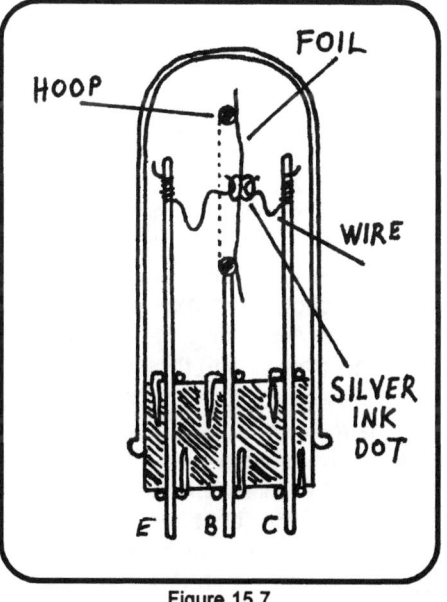

Figure 15.7
Cutaway of the complete Model II

foil, and just as I'd done before, I treated the metal to the modified borax process. The strips oxidized well, and upon testing, the apparent yield of useable PNP structures was found to be much higher.

Needless to say, 0.002-inch foil is quite delicate. It bends at the slightest touch, and while the copper is flexible, the cuprous oxide coating is not. Any deformation of the foil causes the coating to fracture and flake off. Copper-foil PNP structures require additional support.

To address this, the Model II's base electrode was lengthened and fashioned into a hoop. One face of the hoop was moistened with cyanoacrylate (commonly known as "super" glue), and allowed to dry. This is an insulating layer. Glue was applied once more, then the hoop was pressed against the copper foil. The hoop adds strength and support to the foil in much the same way as the rim of a tambourine provides a foundation for its skin.

If you think about it, there is no intrinsic electrical connection between the foil (the transistor's base) and the base electrode. This has to be established by carefully soldering a short strand of fine copper wire from a corner of the foil to the base electrode. See figure 15.6.

The emitter and collector terminals are connected using fine copper wire and silver bearing ink as before. The finished transistor is then inserted into its housing. The Model II can be seen in figure 15.3, while figure 15.7 depicts the Model II's internals.

End Results and Going Further

Before testing for the presence of gain, I tinkered casually with the individual PN junctions to get a feel for their qualitative nature. Using a simple crystal-radio circuit, I replaced the detector with the emitter-base junction of the Model I transistor. I found that the cuprous oxide junction is an exceptional detector, at least for local stations. Subjectively speaking, I found that music and other program material sounded clearer with the cuprous oxide junction than it did when the radio was wired up with some of my galena or pyrite detectors. This may have something to do with the shape of the conductivity curve for this type of junction.

The collector-base junction showed similar promise, as did each of the two junctions in the Model II device. I would argue that, even if you've no interest in building transistors, you might keep cuprous oxide in mind as a substance from which to fashion a good and stable detector.

Back in the chapters on thermionic triodes, I showed how tubes could be set up with variable power supplies and meters to collect data on their response to stimulus. Plotting that data results in a characteristic curve or family of curves that graphically depict the triode's ability to amplify.

Manual data collection can be tedious. Specialized instruments, called curve tracers, are capable of making these measurements quickly and automatically. I have a vintage Tektronix model 575, which is designed to plot data collected from transistors and other semiconducting devices. The curves appear on the 575's screen.

My intent was to connect the Model I device to my 575 and plot its response. I hooked it up, fiddled with the settings on the instrument, and observed some promising shapes, which immediately degenerated to nothing. I checked and double-checked my connections, thinking that a wire had crossed or detached somewhere. In the end, I determined that the collector-base junction had failed, becoming a short circuit. The Model I transistor, such that it was, had exhibited a "useful" life of 10 to 15 seconds.

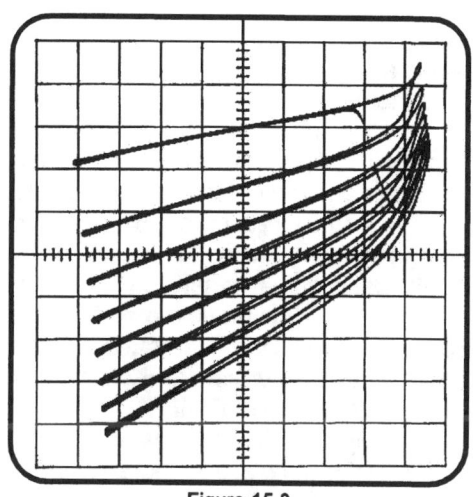

Figure 15.8
Performance curves captured moments before destruction of the collector-base junction

I repeated the experiment with the Model II. Apparent amplifier-like behavior was even more pronounced, but quickly faded. The only record I captured can be seen in figure 15.8. Once again, further tests revealed a short-circuit failure of the collector-base junction.

What I didn't know at the time was that cuprous oxide junctions, when reverse-biased, can generally withstand no more than 6-10 volts. I had set the 575 to apply what I had thought was a conservative value of 20 volts. It appears that this excessive stress drove the collector-base PN junction to the point of destruction.

I would invite you to build a version of the cuprous oxide transistor of your very own. It seems quite possible that such a device, carefully constructed and in the context of appropriate voltages, may demonstrate useable gain.

I have a few parting thoughts on the subject of homebrew transistors: Cuprous oxide seems to disintegrate on prolonged exposure to air. Moisture may play a part in this, or perhaps the cuprous oxide binds with oxygen from the air to degenerate into cupric oxide.

In either case, it pays to protect it as much as possible. One way to shield the oxide is to coat it with a barrier or film of some type. A drop of cyanoacrylate smeared across the surface of the oxide with a toothpick forms a tough, transparent skin that doesn't seem to affect the chemistry of the oxide itself. Perhaps clear nail polish or laquer would achieve the same end, though I can't vouch for this, as I haven't tried them.

It should go without saying that any protective coating shouldn't be applied until all connections have been made to the PNP structure, and their electrical integrity verified.

A curve tracer is an extremely useful instrument to have when doing this type of work. I was fortunate enough to receive my 575 as a gift. Used curve tracers can be found at ham fests and Internet auction sites. New models are still in production, though they can be very expensive. One alternative is to build your own. A simple curve tracer circuit, courtesy of Kazuhiro Sunamura, appears with his permission in Appendix III.

My final thought has to do with the manner in which emitter and collector wires are anchored to the face of the oxide. The silver ink described seems to work well. As near as I can tell, there are no electrical problems with using it. On the other hand, if you'd like to try a different substance with which to bind the wires to the P surfaces, consider metallic indium.

Indium is an extremely soft, almost plastic or putty-like silver metal. Bits of it can be shaved off of an ingot with nothing more than a dull knife. A small particle of it can be mashed against the oxide surface and rubbed and burnished with a blunt toothpick until the metal literally smears and sticks to the oxide. Additional indium can be used to stick an emitter or collector wire to this point of attachment.

Indium is used by the semiconductor industry. Indium pellets can be obtained through chemical supply houses. I purchased a small quantity, more than I will ever need, as surplus through an Internet auction.

Useful References

Coblenz, Abraham, and Owens, Harry L., *Transistors: Theory and Applications*, copyright 1955, McGraw-Hill Book Company, Inc., New York, pp. 16-19, (holes), pp. 45-58, (semiconductor physics), pp. 73-89, (junction transistors)

Conti, Theodore, *Metallic Rectifiers and Crystal Diodes*, copyright 1958, John F. Rider Publisher, Inc., New York, p. 1.6, (accepted reverse-voltage limits for cuprous oxide rectifiers), pp. 2.8-2.13, (construction details for cuprous oxide rectifiers), pp. 3.3-3.6, (the mechanics of conduction in cuprous oxide rectifiers)

Kowsow, Dr. Irving, Editor, *Transistors, A Self-Instructional Programmed Manual,* copyright 1962, Prentice-Hall, Inc., Englewood Cliffs, New Jersey, pp. 98-105, (operation of PNP junction transistors)

Hunter, Lloyd P., Editor, *Handbook of Semiconductor Electronics*, copyright 1956, McGraw Hill Book Company, Inc., New York, pp. 1.3-1.10, (junction and field-effect transistor characteristics)

Kiver, Milton S., *Transistors*, copyright 1962, McGraw-Hill Book Company, Inc., New York, p. 50, (disappearance of point-contact transistors)

Kloeffler, Royce Gerald, *Industrial Electronics and Control*, copyright 1960, John Wiley And Sons, Inc., pp. 45-47, (copper oxide rectifiers, 1,000F forming temperature, inverse voltage limited to 8-10 volts, discovery in 1920)

Krugman, Leonard, *Fundamentals of Transistors*, copyright 1954, John F. Rider Publisher, Inc., New York, pp 1-8, (basic semiconductor physics), pp. 10-15, (operation of junction transistors)

Martin, L.H., and Hill, R.D., *A Manual of Vacuum Practice*, copyright 1947, Melbourne University Press, Australia, reprinted by Lindsay Publications, copyright 1997, ISBN 1-55918-188-5, pp. 87-88, (production of uniform cuprous oxide layer for use in metal to glass seal)

Noon, Walt, *How to Build a Solar Cell That Really Works*, copyright 1990, Lindsay Productions, Bradley, IL, pp. 6-10, (production of cuprous oxide film on a copper substrate)

Pollack, Harvey, *Transistor Theory & Circuits Made Simple*, copyright 1958, American Electronics Co., Mineola, L.I., New York, pp. 15-23, (operation of junction transistors)

Schure, A., *Basic Transistors*, copyright 1961, Hayden Book Company, Rochelle Park, New Jersey, pp. 4-18, (semiconductor physics)

Spenke, Eberhard, *Electronic Semiconductors*, copyright 1958, McGraw-Hill Book Company, New York, p. 73, (cuprous oxide semiconductors), pp. 74-81, (metal to semiconductor contact),

Towers, T.D., and Libes, S., *Semiconductor Circuit Elements*, copyright 1977, Hayden Book Company, Rochelle Park, New Jersey, pp. 1-13, (semiconductor device fundamentals)

Wright, D. A., *Semiconductors*, copyright 1955, Richard Clay And Company, Ltd., Bungay, Suffolk, pp. 41-46, 103 (cuprous oxide, ionic crystals as intrinsic semiconductors)

http://www.jmargolin.com/history/trans.htm
Jed Margolin, "The Road to the Transistor," copyright 1993, 2001, Jed Margolin (origin of copper oxide rectifiers)

Chapter 16
More Semiconductor Ideas

The stable of preceding projects, particularly those involving semiconductors, is by no means exhaustive. In fact, the more reading I did in preparation for this book, the more interesting ideas I unearthed. I tried to make careful notes of the things I found, either because the concepts had sufficient intrinsic value to warrant their inclusion in this book, or because they inspired other ideas that I would surely want to explore in the future. The longer I worked on this, the larger the scope of this book became.

In the end, however, the voice of practicality must be heard. I came realize that, eventually, I'd have to decide when enough was enough, and where it made logical sense to conclude this volume. Editorial decisions are never easy to make because, as in the movie business, somebody's favorite scene always winds up on the cutting-room floor.

This chapter represents a suitable compromise: several sections of basic ideas, processes, and information on the topic of homebrew semiconductors. They may lead to the refinement of previously discussed devices, or the construction of new ones. In any event, I thought them better presented in this format than not at all.

Homebrew Semiconducting Crystals

If you want to experiment with homebrew point-contact transistors, you can use the purified, doped, and crystallized germanium harvested from old crystal diodes. On the other hand, if you'd like to blaze your own trail, you might consider a series of experiments wherein the germanium has instead been replaced with lead sulfide.

Galena, the naturally occurring mineral form of lead sulfide, was once highly regarded as a material from which to build practical point-contact diodes. Personally, I suspect that the first transistor may, in fact, have been born to some anonymous wireless operator who probed his crystals with two whiskers instead of one.

Galena is not scarce by any means, but good-quality galena, electronically speaking, can be hard to find. Galena found in certain mines was known to perform far better than the same mineral from other shafts. I would guess that certain impurities enhance the electrical attributes of the crystal. (Sounds like doping, doesn't it?) Argentiferous galena, for example, seems to have been favored over other types.

If you can't get the galena you want, you might try making your own. The process involves heat, flames, noxious odors, and the handling of lead, but it's not difficult, and not necessarily dangerous if conducted out of doors.

To begin with, you need a source of lead. Fishing sinkers are good candidates. I have recovered sizeable quantities of lead from old wrist and ankle weights. Sometimes discarded vertical blinds contain small pieces of lead, used to add weight. Solder, tire weights, and spent bullets are not pure lead, but lead alloys containing significant amounts of other metals. I'm not necessarily discouraging you from exploiting these materials, but be warned that they may not work as expected.

The first step is to melt the lead in a crucible. I used a clean, 2-inch iron pipe cap as a vessel. The cap was propped up on two bricks, and heated from below with a propane torch. Face protection is a good idea in case the molten lead spatters, and of course, you want to make sure you're set up someplace with plenty of ventilation, as far as possible from anything that's combustible. A corner of the driveway is a nice, safe place.

Sulfur must be added to the lead. Sulfur is available as a yellow powder. You can get it through a chemical supply house, you may find it at the pharmacy, or you can do as I did and purchase it at your local lawn and garden shop. Sulfur is used to dust rose bushes.

I used an old teaspoon to add sulfur to the lead, while I stirred the mixture with a charred (carbonized) stick. Be warned, the sulfur will probably burst into flames, releasing choking fumes. Keep the bulk container of sulfur away from your propane torch, and don't breath the smoke that boils out of the crucible. Keep adding sulfur and stirring the mix until the lead thickens, clumps and congeals into a black mass.

I removed the heat, allowed the crucible to cool, and then chiseled out the gunk that I'd created. I pulverized this matter with a hammer, and ground it into particles as fine as I could.

Next, I procured a 1-ounce piece of silver bullion from a local coin dealer. At the time, an ounce of 0.999 percent pure silver ran about six dollars. Using a clean file, I filed an edge of the bullion to produce a small pile of silver filings.

Starting with a measured quantity of the powdered gunk created earlier, I added additional sulfur and then silver filings to create a new mixture with the ratio of 100 parts gunk to 10 parts sulfur and 3 parts silver, by weight.

The triple mix of powders was dumped into a fresh pipe cap and heated on a small propane burner to a red heat. At the conclusion of the process, black crystals of synthetic argentiferous galena were found in the cap. I chose the best of these and mounted them in slugs of Wood's metal. Wood's metal, by the way, is still available from the distributors of machinist's supplies, under the trade name Cerrobend.

An Unusual Transistor Material

When I originally introduced the topic of semiconductors, I made passing reference to the experiments of Robert Pohl and Rudolph Hilsch. In their 1938 paper entitled "Control of Electric Currents With a Three-Electrode Crystal and a Model of a Barrier Layer," they describe an amplifying device based on a potassium bromide crystal.

While I was unable to locate explicit details, Eckert and Schubert's book, *Crystals, Electrons, Transistors - From Scholar's Study to Industrial Research*, offers clues that an enterprising experimenter might be able to run with.

Potassium bromide, formula KBr, can be purchased as an unremarkable white powder. Among other things, it apparently finds application in the field of veterinary medicine for the control of seizures. Grown into crystals, KBr becomes clear, making it useful for specialized optical applications as well.

A bromide crystal, mounted between suitable electrodes, is said to exhibit interesting properties. See figure 16.1. On application of an electric current, trapped electrons cause the formation of a blue cloud that migrates through the crystal. Note that the positive terminal is essentially a plate, while the negative terminal is point-contact. Regrettably, I have no information on the dimensions of the crystal, the

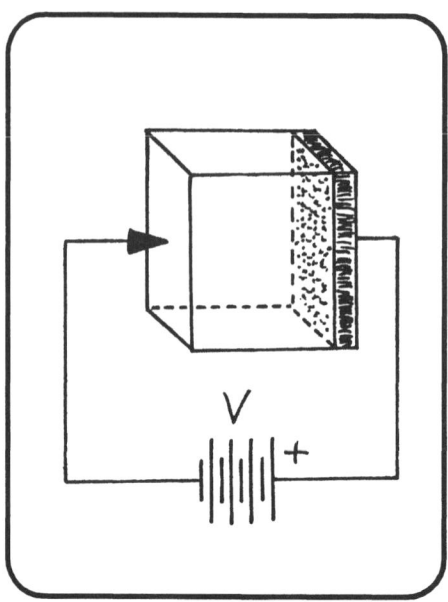

Figure 16.1
Applying a voltage to a KBr crystal

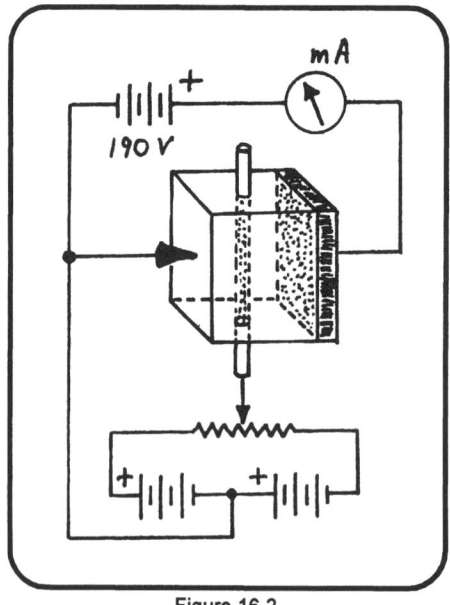

Figure 16.2
Pohl and Hilsch's KBr "amplifier"

composition of the electrodes, or the applied voltage required to produce the effect.

Figure 16.2 depicts what I know of Pohl and Hilsch's instrument of amplification. The body of the device was a KBr crystal with a point-contact at one end, and a plate electrode at the other. These can be likened to the thermionic triode's filament and plate, respectively. An excerpt from Hilsch's laboratory notebook, reproduced in Eckert/Schubert's book, seems to imply that the plate voltage on the crystal was on the order of 190 volts.

Piercing the crystal is a thin wire or rod that acts as a grid to control the flow of current through the crystal. According to Hilsch's notes, the voltages applied to the crystal's "grid" ranged from -30 to +40 volts, with respect to the point-contact. What is not evident is how the wire became embedded in the crystal. Perhaps the crystal was actually grown around it. Who knows. In any event, the current gain claimed for this device is on the order of 100 to 1.

Reportedly, this amplifying device did not

find practical application because of the "...low drift velocity of the electrons." If there was a significant delay in the propagation of electrons through the crystal, it would certainly make it unsuitable for the amplification of high-frequency signals like radio waves. If the movement of electrons is extremely slow, even audio signals would be impaired. On the flip side, there are other applications for direct-current amplifiers. Moreover, it is claimed that the operation of the device can actually be witnessed as the movement and diffusion of colors through the crystal, ample justification in itself for building such an instrument!

An interesting side note: Hilsch's notebook records a temperature "T" of "420 degrees." No units are specified, nor is there any clarification as to what, precisely, has been heated to this temperature. I presume it is the operating temperature of the crystal itself.

I would guess this value expresses degrees Celsius, or possibly, degrees Kelvin. Either way, is appears the KBr crystal has to be pretty hot to work as advertised. (Its melting point, by the way, is 730 degrees C.)

The Field-Effect Transistor

The Weagant tube, described back in Chapter 12, was a variation of the classic thermionic triode. If you recall, it consisted of a filament that emitted electrons, and a plate that collected them. Both were confined to a narrow, evacuated tube. The flow of electrons between the two was controlled not with an invasive grid, but through variations in a static field applied from the *exterior* of the tube with a metal sleeve.

An analogous device, called a *field-effect transistor* or FET, exists in the semiconductor world. FETs come in number of variations, but generally speaking they're composed of a narrow strip of N-type or P-type semiconducting material, called a *channel*. One end of the channel becomes the *source* electrode (analogous to the triode's filament) while the other is called the *drain* (analogous to the plate). Layered on top of the channel, but insulated from it, is an electrode called the gate. It behaves much like the control electrode wrapped around the waist of the Weagant tube. Changes in the charge on the gate act to constrict the flow of charge carriers through the channel.

You might think the FET too exotic a device to homebrew. Once again, research revealed that somebody, indeed, had done precisely this. R. Baker's experimentation with FET-like structures was documented in the "Amateur Scientist" column in the June, 1970 issue of *Scientific American*.

To be brief, Baker's FETs are based on semiconducting channels composed of thin films deposited on an inert substrate. The films are created when special mixtures of chemicals are applied to glass. If you're interested in replicating the work, it's best to secure a copy of the original article. I'd recommend the purchase of the "Amateur Scientist" CD-ROM, mentioned in a previous chapter. In any event, the construction of a sample FET can be summarized as follows:

Baker begins with a substrate composed of glass, cut to appropriate size with a diamond point. The pieces are soaked for three days in an acid solution composed of 1 part nitric acid to 12 parts distilled water. This is said to leach out calcium and sodium ions. Presumably, the glass is then removed and rinsed.

The film material consists of 500 milliliters of distilled water with a .01 M (molar) concentration of thiourea and a .01 M concentration of cadmium chloride.

A piece of substrate material is placed in a beaker and immersed in the film solution. Concentrated ammonium hydroxide is added until the mixture turns faintly cloudy and then clears. The beaker is then placed in a double boiler and slowly heated to boil for 15 minutes. The film solution is supposed to turn yellow-orange, signaling the precipitation of cadmium sulfide.

The glass substrate, which has become coated in the process, is bathed and rinsed in distilled water. The sulfide coating can be doubled by repeating the above process.

Next, the glass is heated to 500 degrees C for 30 minutes. With the film firmly set, Baker cut the coated substrate into smaller pieces measuring 1/4 by 1/2 inch.

Construction of Baker's FET is mostly mechanical from that point on. A source and drain electrode are applied as thin strips of indium burnished onto the sulfide film. The film is coated with an insulator (vinyl cement), followed by the application of a "silver paste." The conductive paste forms the FET's gate. The entire structure, with the exception of the terminals, is coated and sealed with silicone rubber. See figure 16.3.

The "Amateur Scientist" article goes on to describe simple electrical tests showing the FET's ability to act as an instrument of amplification. Depending upon the charge applied to the gate, the source-to-drain current can be made to vary from as little as 10 to 50 microamps, with an applied voltage of 9 volts.

I have not yet had the opportunity to try Baker's process, but a few comments come to mind as I read about it.

Many of the chemicals specified are very toxic. The fumes produced by boiling these substances or spraying them onto heated glass are undoubtedly bad news. Personally, I would only attempt this type of work outdoors. Even then, I think I'd invest in a protective mask of some sort.

Some of the chemicals are relatively exotic, including our old friend, nitric acid. The comments I made in the last chapter about the procurement of chemicals for experimentation applies here as well.

In case you're wondering, the term "molar" is an expression of the number of molecules of a substance dissolved in a given volume of solvent. In this case, we're speaking about various chemicals in a given quantity of water. Knowing the desired molar concentration, and the amount of solvent, it's a fairly simple matter to compute the number of grams of each chemical that must be added. Any introductory chemistry book will show you how.

Despite some of its drawbacks, Baker's thin film approach to FET construction is a nifty one. It prompted me to consider other films and how they might be produced.

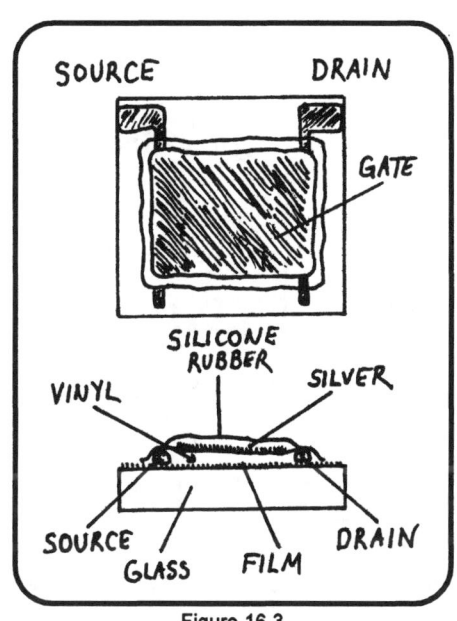

Figure 16.3
Construction of Baker's homebrew FET

A 1931 reference on mirror making, produced by the Bureau Of Standards, mentions a technique for applying a lead sulfide film to glass for the purpose of creating "black mirrors." The chemistry is similar to Baker's process, the heart of which is "...1 g of thiourea in 50 to 75 ml of water," to which is added "... 50 to 75 ml of a dilute solution of lead acetate and finally 25 ml of dilute potassium hydroxide or ammonia solution, mixing constantly." The author notes that the sulfide films can be used as "... electric resistances of small dimensions with

Page 277

a relatively high resistance, and they are, perhaps useful for other purposes."

Why not try lead sulfide as the basis for a Baker-type FET? We're already familiar with its semiconducting properties in the form of galena.

Finally, delicate metallic films can be applied to glass through a technique called *sputtering*. To sputter, you immerse a target metal in a glow discharge produced inside of an evacuated chamber. The movement of charged particles causes bits of the target to be eroded away and the fragments deposited on nearby surfaces.

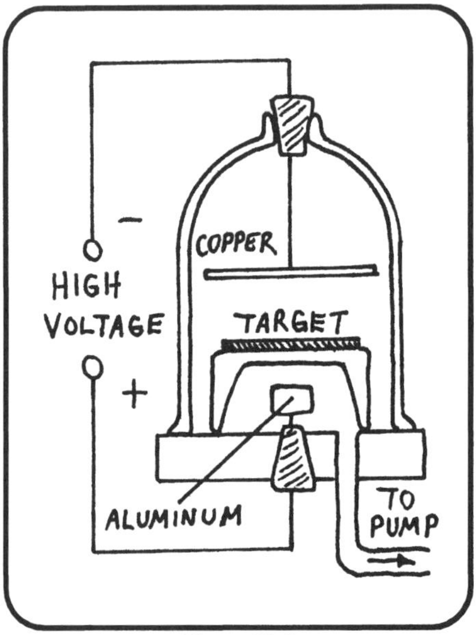

To accomplish this, you'll need a glass vessel, a vacuum pump, and a source of high voltage. (Note that, if you successfully built the glow tube described a few chapters back, you basically have all of the equipment you need.)

Like the glow tube, the sputtering chamber has two electrodes. See figure 16.4. The negative electrode is made of the substance with which you intend to sputter. Lets assume, for the moment, that it's copper. The composition of the positive terminal is pretty much up to you. Aluminum is as good as any material. An insulated table, or platform, is installed to keep the item you wish to sputter in close proximity to the target.

Figure 16.4
Using a glow discharge to sputter metals onto a glass substrate

Once the internal pressure of the chamber has been reduced to 1 millitorr or so, you can apply the high voltage and watch it glow. Slowly, tiny particles of copper will appear, both on the surface of your substrate, and on the walls of the chamber in the vicinity of the cathode.

While I haven't built a chamber for the expressed purpose of sputtering, I did observe, first hand, the effects of sputtering in my earlier glow tubes. After prolonged operation (several hours) I noticed a coating of copper on the interior of the glass bulb near the negative terminal. While thin enough to remain semitransparent, the coating was nonetheless substantial enough to actually conduct electricity!

Negative-Resistance Devices

If a large pipe carrying a certain flow of water is fitted with a narrow restriction, one will immediately notice a drop in the pressure of the water exiting the pipe. The drop is the result of the losses incurred as the water fights its way through the restriction.

The same is true for an electric current passing through a resistance. In this latter case, it is the voltage that drops as the current passes through the resistor. A predictable outcome of Ohm's law, this sort of loss is associated with all conductive materials to a greater or lesser degree.

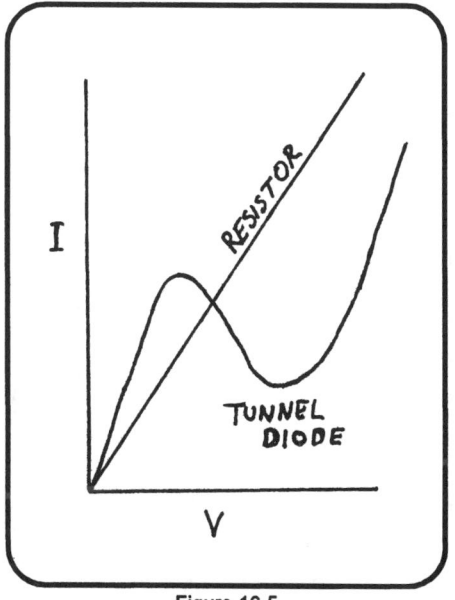

Figure 16.5
Behavior of a negative-resistance device versus an ordinary resistor

Suppose for a second, that it was possible to fabricate a material that had an anti, or negative, resistance. The loss in voltage previously observed in the circuit, would instead appear as an *increase* in voltage, representing a kind of gain.

It turns out that several devices exhibit the seemingly magical property of *negative resistance*, at least over a limited set of conditions.

In October of 1957, Leo Esaki announced the results of studies he had conducted into the "tunneling" effect of electrons in thin semiconductor junctions. The practical result of his work was the invention of the *tunnel* diode. As instruments of amplification, tunnel diodes offer the ability to amplify and oscillate at high frequencies.

They are also said to be more resistant than ordinary transistors to the destructive effects of exposure to nuclear radiation. The graph in figure 16.5 depicts the relative difference between the electrical behavior of a regular resistor verus the tunnel diode.

As you may have already guessed, it's possible to fabricate a tunnel diode–like device, and happily, no exotic materials are required. Best of all, the resulting device is of sufficient quality to be useable in practical circuitry.

Nyle Steiner, an amateur science wizard that I've mentioned before, has fabricated simple negative-resistance devices from strips of galvanized steel, which are then probed, point-contact–style, with a fine wire.

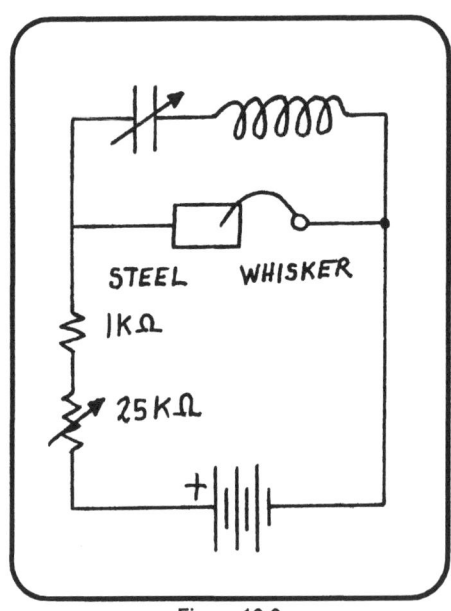

Figure 16.6
Steiner's negative-resistance
broadcast oscillator

Galvanized steel is coated in zinc. Steiner harvests his raw material from ventilation duct work. Using pliers to hold the metal, he heats a 1/8-inch wide strip in the flame of a propane torch until the strip glows and begins to cast off sparks. When removed from the flame, and as the strip cools, dark, black spots will begin to pepper the metal's surface.

Mr. Steiner suggests that the dots are composed of zinc ferrite, formula $ZnFe_2O_4$. Furthermore, he cites the *Handbook of Chemistry and Physics* as describing zinc ferrite as a "black material." Having done a bit of peripheral reading on the subject, I've concluded that it's a reasonable assertion.

Figure 16.6 shows the schematic for a simple radio frequency oscillator based on the galvanized metal device. The circuit can be made to oscillate from audio frequencies up to an apparent ceiling of about 2 MHz. This means it has the potential capability to transmit to a nearby AM radio.

Mr. Steiner has had similar success with galvanized nails, steel-to-aluminum contacts, and probes applied to iron pyrites, silicon, and to the surface of zinc chromate–plated nuts. His experiments are well documented on his Web site, and he's written several articles on the subject for the Xtal Set Society's Newsletter.

While I have not tried to replicate Mr. Steiner's experiments, it occurs to me that the homebrew transistor housing described in Chapter 14 would make an excellent enclosure for a more permanent model of Steiner's homebrew tunnel diodes.

References

Bobrow, Jerry, (editor), *Cliffs Quick Review Chemistry*, copyright 1993, Cliffs Notes Incorporated, Lincoln, Nebraska, pp. 81-92 (solutions)

Carlson, Shawn (editor), *The Amateur Scientist - The Complete Collection on CD-ROM*, copyright 2000, The Tinker's Guild, Menlo Park, California, (multiple references to "Amateur Scientist" articles)

Carrol, John M. (editor), *Tunnel-Diode and Semiconductor Circuits*, copyright 1963, McGraw-Hill Book Company, Inc.,New York, p. 10, (Esaki and the origin of the tunnel-diode), pp. 10-18, (tunnel diode properties and applications),

Eckert, Michael, and Schubert, Helmut, *Crystals, Electrons, Transistors*, copyright 1990, American Institute of Physics, New York, pp. 106-109, (crystal-triode experiments with potassium bromide)

Gardner, I.C., and Case, F.A., *The Making of Mirrors by the Deposition of Metal on Glass*, copyright 1931, Bureau of Standards, Washington, D.C., pp. 13-14 (deposition of lead sulfide on glass), p. 14, (cathode sputtering)

Hunter, Lloyd P., Editor, *Handbook of Semiconductor Electronics*, copyright 1956, McGraw Hill Book Company, Inc., New York, pp. 1.3-1.10, (junction and field-effect transistor characteristics)

Steiner, Nyle, "Crystal Set Detector With Gain Using Negative Resistance," *The Xtal Set Society Newsletter*, Volume 12, Number 4, Hewes, Rebecca, (editor), copyright 2002.

Wailes, Raymond, Editor, *Manual of Formulas - Recipes, Methods, and Secret Processes*, Popular Science Publishing Co., Inc., copyright 1932, reprinted by Lindsay Publications, Bradley, Illinois, pp. 132-133, (a process for producing 'radio crystals')

http://home.earthlink.net/~lenyr/
Steiner, Nyle, "Spark, Bang, Buzz and Other Good Stuff," copyright ?, last updated March, 2001, (negative-resistance amplifiers based on heated, galvanized metals)

Chapter 17
Parting Thoughts

Most of the nonfiction books I've read over the years flow according to a familiar and time-tested convention. Usually, things start with an introduction. This is followed by the presentation of information as part of a larger body of text. Finally, a conclusion of some sort wraps everything that preceded it into a neat package, suitable for intellectual consumption. It's over and done.

The trouble with concluding paragraphs is that they are invariably seen as the end of the reader's journey. In stark contrast, the final pages of *Instruments of Amplification* clearly represent a beginning.

Nobody can compress into a few hundred pages everything there is to know about the topics we've just explored. There is simply too much to know. On the other hand, there are literally thousands of other science books in circulation—physics, chemistry, and engineering—that address these topics in greater detail. Hunting for more information and understanding it is an expedition in its own right, limited only by your curiosity and willingness to learn something new.

Then, there is the matter of the physical construction of the instruments described here. There is something wonderful about converting a thought into tangible object. The process can teach new skills and expand one's mind. It's also amazing how aware one becomes of new tools, technologies, techniques, and materials.

Nor is there any obligation on your part to follow a beaten path. Who says your finished devices have to look or function like mine? Your ideas may take things in a completely different direction. Perhaps the instruments you build will work even better. In fact, it's not impossible for an amateur to discover something totally new.

You see, once you dive into the pool, where you choose to swim is entirely up to you.

Appendix I
High-Voltage Supplies

High-voltage power supplies are not only interesting, they're useful. There are a variety of ways to generate high voltages, including transformers, induction coils, and "flyback" transformers.

Figure A1.1 shows how to wire up an old ignition coil to generate a continuous stream of sparks. The circuit is somewhat more sophisticated than the simple hookup described in Chapter 10. Instead of using a buzzer or small motor to interrupt the primary current feeding the coil, this design uses a power transistor to do the switching.

The transistor is triggered on and off by a signal generated by a common and inexpensive timer chip—the ubiquitous 555. The transistor, a 2N3055, will no doubt get hot during operation, and should be mounted on a suitable heat sink.

Note that A1.1 is offered by way of reference only, as I have not constructed the circuit myself.

The circuit in figure A1.2, based on the flyback transformer found in television sets, is both simple and generic. Composed of little more than a couple of power transistors and two resistors, it can generate extreme voltages.

Flyback transformers, depending upon the equipment from which they are harvested, may differ in their characteristics and construction. One experimenter claims that only those transformers salvaged from black and white televisions will work properly.

Some years back, I constructed and experimented with several flyback circuits of the type shown, and with great success. I can't say for certain what sort of junk my flybacks were harvested from, but I do recall tinkering with color television parts at the time. I would not rule out flyback transformers from dead computer monitors, either.

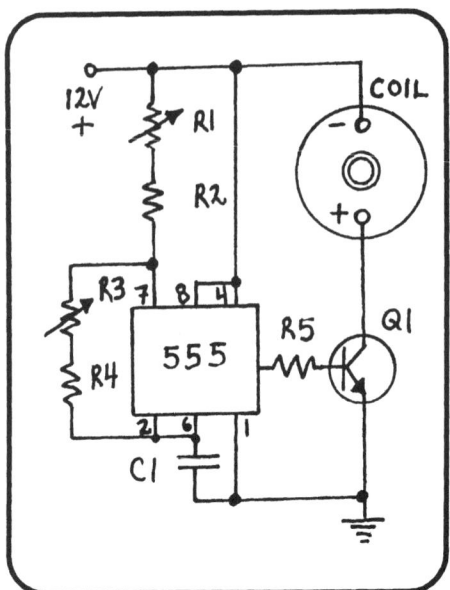

Figure A1.1
A high-voltage supply using an ignition coil

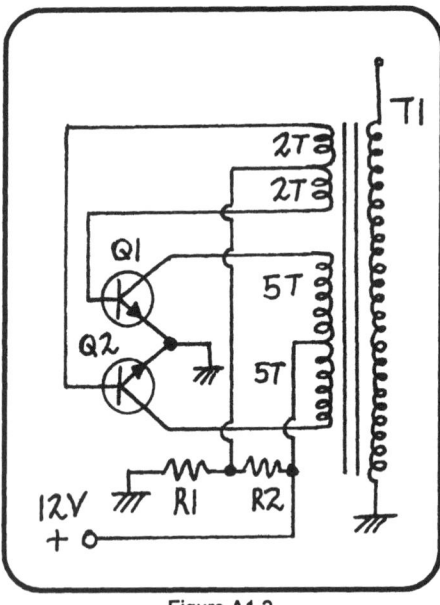

Figure A1.2
A high voltage supply using a television flyback

A number of simple windings must be added to the core. The first, a driver coil, is composed of 10 turns of 14- or 16-gauge wire. The coil is center tapped.

The next coil is a feedback winding. It consists of four turns of 22-gauge wire. It is also center tapped. If the circuit fails to self-start, try swapping the outer leads of the feedback coil.

As before, the transistors will generate a good bit of heat during operation, and must be mounted to heat sinks to prevent self-destruction.

In this case, the output is a high-voltage, high-frequency, alternating current. As long as you're scrounging flybacks out of old sets, try to salvage the high-voltage rectifier (diode) as well. It can be connected to the high-voltage lead of the flyback to give you direct current. This type of supply is excellent for sputtering applications.

Given the nature of these devices, a few warnings are in order. The circuits above are intended to be powered by batteries or some other source of direct current, producing 12 volts or less. Do not attempt to "increase the power" by connecting these to some other energy source.

While I don't think these supplies would likely electrocute any healthy adult, they demand respect just the same. A person with a weak heart or other frailties might, indeed, be killed by contact with the output terminals. These supplies are more than capable of inducing pain, causing unpleasant muscle contractions, and serious burns. They can easily ignite any combustibles in the path of their sparks. In fact, a couple of decades ago, I would amuse curious friends by demonstrating how a flyback supply could generate an arc capable of punching through presumed insulators like cardboard, plastic, and even thin sheets of glass!

Additional examples of circuits of this type can be found all over the Internet. An good example of this is *Snock's World of High Voltage*, a series of web pages found at:

http://www.geocities.com/CapeCanaveral/Lab/5322/index.html

The circuit in figure A1.1, by the way, was adapted from Snock's 15-40 kV ignition-coil circuit.

Needless to say, power supplies of this type are intended for mature and responsible experimenters. Don't get bit!

Parts List for Figure A1.1

Part	Description
C1	0.1-uF capacitor
Q1	2N3055 power transistor
R1,R3	10-k potentiometer
R2,R4	1-k resistor
R5	100-ohm resistor

Parts List for Figure A1.2

Part	Description
Q1,Q2	2N3055 power transistor
R1	27-ohm, 10-watt resistor
R2	220-ohm, 10-watt resistor
T1	Flyback transformer salvaged from television set (Additional windings must be added to the transformer in order to use it... see text.)

Appendix II
A Variable Plate Supply

Plate voltages for homemade thermionic triodes can be supplied with a half-dozen or more 9-volt batteries, wired together in series. Unfortunately, batteries wear out.

I saved myself a small fortune in replacement costs by building a power supply. My supply saves money on batteries, is more reliable, is fully adjustable, and offers isolation from potentially dangerous house mains.

My power supply can be seen in figure A2.1, its schematic appears in figure A2.2. The "variac" T1 is an adjustable form of transformer. I found mine in some electronic scrap, but there are numerous models still in production.

The isolation transformer T2 protects me from accidental contact with home power mains, even though the supply plugs into the wall. This is particularly important when you're playing with radio circuits that drive headphones.

The transformers produce alternating current. This must be converted to direct current with a rectifier bridge and filtered with capacitors (condensers). The latter were salvaged from the power supply of a discarded personal computer. The choke, L1, is actually the 24-volt secondary of a small filament transformer.

I built the circuitry into a wooden file box. I added an old meter so that I could monitor the supply's output.

Figure A2.1
A home built plate supply

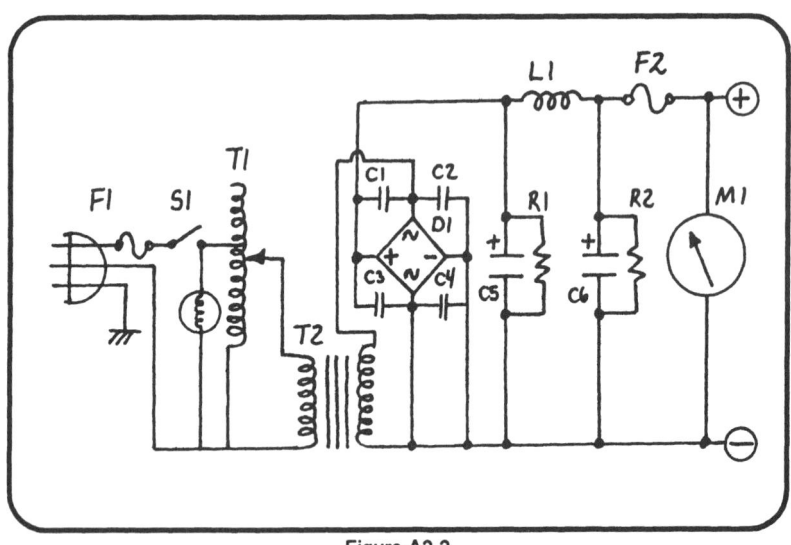

Figure A2.2
Schematic for the plate supply

Parts List for Figure A2.2

Part	Description
C1, C2, C3, C4	0.01-uF disk capacitor
C5, C6	330-uF electrolytic capacitor, salvaged from computer power supply
D1	Full wave diode bridge, salvaged from computer power supply
F1, F2	Fuse, 0.5-amp glass cartridge
L1	Choke (use the secondary of a small 24-volt filament or bell transformer)
M1	0-150-volt D.C. meter movement
R1, R2	47-k bleeder resistors
T1	Variable transformer, Ohmite VT1R5 or similar (120 volt input, 0-135 volt output)
T2	115-volt to 115-volt isolation transformer, 15 watts

Appendix III
A Simple Curve Tracer

A curve tracer is a useful instrument to own if you're building your own semiconductors. Its display can tell you, at a moment's glance, how your device is performing. Unfortunately, new ones can be pricey.

My own curve tracer is a Tektronix 575. It's a very old, vacuum-tube–based instrument, but it still works like a champ. I was fortunate to have received it as a gift.

In searching the Web for affordable alternatives, I was delighted to discover a circuit published by Kazuhiro Sunamura. It plots performance curves on a common oscilloscope. Figure A3.1 shows the schematic, reproduced here with his permission.

The transistor at the far right in the schematic is the transistor whose characteristics are to be plotted. An oscilloscope, set up for X-Y display, is connected to the test points marked "X" and "Y". Display ground can be taken from the low side of R8.

As is, the circuit can only be used with NPN transistors, though other experimenters report having modified and enhanced the design to accommodate the testing of both NPN and PNP devices.

More details can be found on Mr. Sunamura's Web site at :

http://www.intio.or.jp/jf10zl/trct.htm

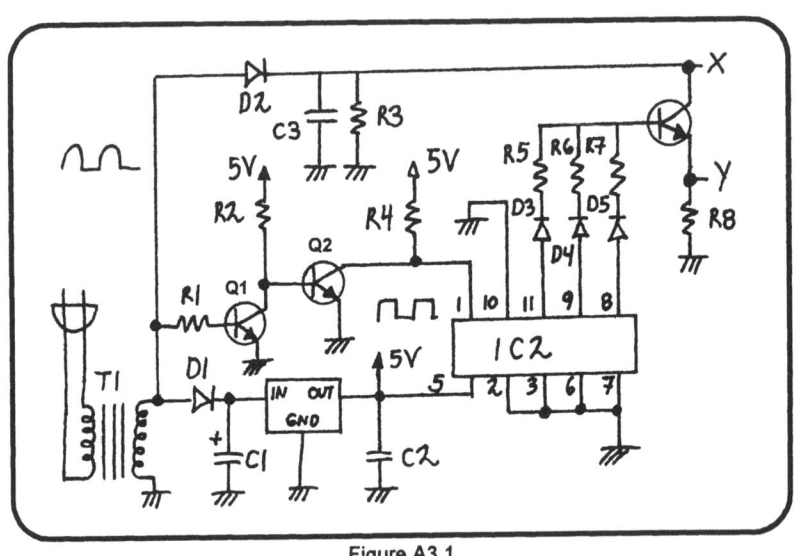

Figure A3.1
Schematic for a simple, inexpensive
curve tracer

Parts List for Figure A3.1

Part	Description
C1	470-uF electrolytic capacitor
C2	0.01-uF capacitor
C3	0.008-uF capacitor
D1, D2	1S1588 diode
D3, D4, D5	Schottky diode (part number not specificed)
IC2	7490 TTL decade counter ic
IC1	7805 three-terminal voltage regulator ic
R1	10-k resistor
R2, R3	15-k resistor
R4	1-k resistor
R5	39-k resistor
R6	82-k resistor
R7	169-k resistor
R8	10-ohm resistor
Q1,Q2	2SC1815
T1	Power transformer, 120 volts to 16 volts

Appendix IV
Specifications for a
Point-Contact Transistor

The following data was taken from Raytheon specification document Rev. 2-CS-2523, dated April 21, 1949.

Device:
Transistor CK703

Application:
Medium-gain crystal triode intended for use as an amplifier for frequencies ranging from audio to video

Maximum Ratings:

Collector Voltage:	-70 volts
Collector Current:	4 milliamps
Collector Dissipation:	200 milliwatts
Emitter Current:	10 milliamps

Typical Characteristics:

Collector Voltage:	-30 volts
Emitter Voltage:	0.2 volts
Collector Current:	2 milliamps
Emitter Current:	0.75 milliamps
Transconductance:	5,000 micromhos
Collector Impedance:	10,000
Emitter Impedance:	500 ohms
Average Power Output (50 microwatts input):	2 milliwatts
Average Power Gain:	16 decibels